10/94

The Changing World of Weather

The Changing World of Weather

Clive Carpenter

Facts On File
New York • Oxford

Editor: Beatrice Frei
Design and Layout: Eric Drewery
Picture Editor: P. Alexander Goldberg
Additional picture research: James M. Clift
Special thanks to: Bernard Green, Marie Hélène Cambos,
Catherine Terk and Anne Jones. Also: National Meteorological
Library, Spectrum Colour Library (London), Gamma Presse
(Paris), Images Colour Library (London), Explorer Archive
(Paris), Mary Evans (London), Telegraph Colour Library (London).

Published in Great Britain by Guinness Publishing Ltd,
33 London Road, Enfield, Middlesex

Facts on File, Inc.
460 Park Avenue South
New York NY 10016

Library of Congress Cataloging in Publication Data
Carpenter, Clive.
  The changing world of weather/Clive Carpenter.
    p. cm.
  ISBN 0–8180–2521–5
  1. Climatology. 2. Weather. 3. Climatic changes. I. Title.
OC981.C35 1991   90–46709
661.6––dc.20   CIP

Composition by Ace Filmsetting Ltd, Frome, Somerset
Printed in Hong Kong by Imago

10 9 8 7 6 5 4 3 2 1

Contents

For Gill

Picture captions for chapter openers:

P. 8/9   A plague of crickets followed the drought in Senegal, Africa. (Gamma)

P. 28/29   The exposed Channel coast of Belgium was battered in the storms of January and February 1990. (Gamma)

P. 44/45   Sunflowers growing in southern England. A series of warm summers has encouraged agricultural diversification. (Spectrum)

P. 56/57   Morteratsch Glacier in Engadin, Graubünden, Switzerland. The advance and retreat of Alpine glaciers with climatic change is well documented. (The Photo Source)

P. 64/65   The European settlement of Greenland, 1000 years ago, coincided with a mild climatic period.

P. 76/77   A cumulonimbus cloud bursts through the surrounding layer of altocumulus. (Spectum Colour Library)

P. 96/97   Polythene, used to suppress evaporation on agricultural land. Global warming could greatly increase evaporation in many marginal areas and endanger existing farming economies. (Gamma)

P. 112/113   Members of the Yanomami tribe. Deforestation in Brazil threatens the livelihood of this and other tribes, as well as posing a global threat to established climatic patterns. (Gamma)

P. 128/129   The Thames Flood Barrier, Woolwich, London, would, according to some scientists, be unable to cope with the scale of sea level changes anticipated as a result of global warming. (Spectrum)

P. 144/145   Tundra landscape in spring. Large quantities of the greenhouse gas methane trapped in the tundra soils could be released by global warming. (Spectrum)

P. 164/165   Scientist with weather radar monitors. Increasingly sophisticated technology forecasts short-term weather patterns and long-term climatic variations. (SEFA)

Introduction

Over the past decade abnormal weather conditions have made headline news in many parts of the world. Storms of unusual ferocity, prolonged periods of drought and devastating floods have fuelled public concern that 'something is happening', that a climatic change is under way. But that concern is not universal. Some regions of the world have always been subject to extremes of climate and some Americans have wondered why Europeans became so alarmed by the great storms of 1987 and 1990 and the extended drought of 1989–90 when conditions such as these were not an unusual experience in parts of the United States.

This book describes the freak weather conditions that have been experienced not only in Europe but also in other parts of the world, and suggests the reasons why the weather seems to be changing. It sets the recent unusual conditions in context by an examination of the climate of past periods and highlights similar storms and droughts in other centuries as well as dramatic climate changes that have caused the depopulation of whole regions.

The reasons for the present distribution of climatic belts round the world and the factors that govern our weather are described, among them the human factors that could lead to a climatic change. Prime amongst these are the emission of greenhouse gases and the felling of the tropical rain forests. Many people are becoming seriously worried about the possibility of global warming through the greenhouse effect. This book details the greenhouse theories and looks at some of the results that have been forecast, including flooding, desertification, and major changes in settlement and agricultural practices.

The greenhouse effect is seen by many as the culprit for the perceived changes but there are alternative explanations. Ocean currents, whose major climatic role is only now beginning to be realized, and variations in solar radiation are the most important of these and both are highlighted in the final chapter.

The Changing World of Weather does not, though, look only at the grand scale of climatic change. It details the essential background knowledge that helps us to understand the day-to-day features of weather—clouds, weather maps, raindrops, wind speeds and much more besides. Above all, it sets out to show that the planet and its atmosphere is not a static cell, unchanging and constantly predictable, but rather a vibrant weather 'machine' that is ever active, ever changing and endlessly fascinating.

1 What is happening to the weather?

In February 1988, snow fell in the Syrian Desert. In February 1990, Stockholm Harbour, which would normally have been choked with ice, had open water as the city enjoyed the warmest February since records were first kept in Sweden in 1700. In February 1989, British holiday-makers were startled to see snow-flakes in the Canary Islands when the weather they had left at home was so mild that the spring flowers were in bloom five to six weeks early.

From all round the world in the late 1980s came reports of freak weather conditions—of extremes of heat and drought, especially mild winters, unusually heavy rain and snowfall, and storms of frightening intensity. These extremes continued into the 1990s with severe gales and then drought in the United Kingdom, snow in Florida and no snow in Alpine skiing resorts.

Occasional extremes of heat and cold and of precipitation are not in themselves exceptional. Any meteorological officer will tell you that weather records are being broken all the time. But, over the last five years, it has sometimes seemed that familiar weather patterns are no longer to be relied upon. Some observers believe that we are witnessing the beginning of a profound climatic change, but most scientists hold that it is too soon to suggest anything more serious than a series of unusual climatic variations that just *might* be the harbingers of a long-term fluctuation.

We have all read that as a result of various human activities—the destruction of the great tropical rain forests, the emission of 'greenhouse' gases, for example, by car exhausts and so on—the Earth is getting warmer. It is assumed that global warming is a reality and, as a consequence, so is climatic change.

The truth is uncertain. Although the bulk of evidence available shows that our planet is getting warmer, the amount and the rate of temperature increase are both disputed. Nor has it been proved that the cause of the increase in temperature is the 'greenhouse effect'. What we do know is that the weather, in many places, does not appear to be sticking to what we thought were the established climatic rules.

Cannes, on the French Riviera, owes much of its popularity to the mildness of its climate in winter. Encouraged by the example of the Lord Chancellor, Lord Brougham, who patronized Cannes every winter for 34 years, large numbers of the English Victorian aristocracy flocked to the town to escape the fog and the cold of London. Cannes grew into an elegant and fashionable resort, famous for its winter season. In the 1980s, weather patterns seemed to be overturned as here at Cannes, blanketed in snow. (Gamma)

DROUGHT IN THE SAHEL

Few regions illustrate more clearly the substantial variations in climate that have been experienced in the past decades than the Sahel, a wide belt that stretches across Africa from Senegal in the West to the Sudan in the East. Since 1968 the Sahel has suffered a prolonged—but discontinued—drought, but it is too early to tell if this represents a major long-term change in climate or whether it is only a temporary variation.

The Sahel was one of the world's great natural pasturelands. Across its relatively flat surface and open horizons were seemingly endless expanses of low grasses dotted by single acacia, baobab and other deciduous trees, or by patches of scrubby thorns. In Senegal, Mauritania, Mali, Burkina Faso, Niger, Nigeria, Chad and the Sudan, the vast grasslands of the Sahel provided forage for a variety of herds and flocks—camels, cattle, sheep and goats.

The balance between limited natural resources and human activities was always difficult to maintain. Husbandry had to be practised with great care as there was the ever constant danger of overgrazing and overstocking. If the discontinuous vegetation cover was destroyed, desert conditions could so easily gain ground.

By the 1960s overpopulation, and the accompanying overuse of the land, was already beginning to take a toll on the Sahel. But although, on average, eight months of the year were dry, the limited amount of rain that did fall in the Sahel—in some places 100–200 mm (4in to 8 in), but in other places considerably higher—was relatively reliable. Usually the rains came in a short 'wet' season between June and August, and their sudden appearance brought flooding to the Rivers Niger and Senegal and to Lake Chad, which had been shrinking. This small amount of precipitation was enough to bring new growth to the poor pastures of the region.

A severe drought began in the Sahel in 1968. In the poor land-locked republic of Mali, crops virtually disappeared. Some two thirds of Mali's livestock died in the drought, and the nation's meagre rainfall totals decreased to a level at which human life was threatened. By 1974 large areas of Mali's natural pastures had been lost as the Sahara Desert had advanced in

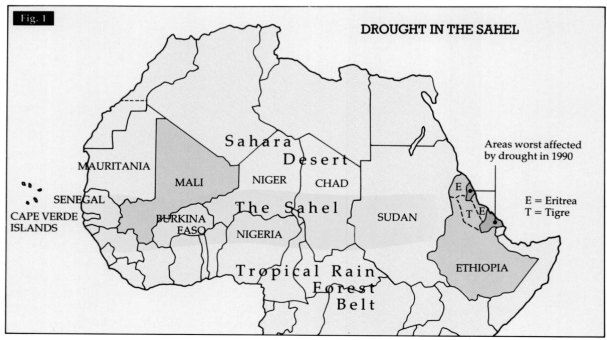

DROUGHT IN THE SAHEL

Sahara Desert

The Sahel

Tropical Rain Forest Belt

MAURITANIA

MALI

NIGER

CHAD

SENEGAL

CAPE VERDE ISLANDS

BURKINA FASO

NIGERIA

SUDAN

ETHIOPIA

Areas worst affected by drought in 1990

E = Eritrea
T = Tigre

Countries highlighted in brown are featured in the text.

Left Attempts to convert the tropical grasslands of the Sahel to agriculture have met with mixed success. The variability of rainfall has made crop production erratic. The prolonged drought of the 1980s devastated their lives. In Ethiopia and in Sudan millions of tonnes of emergency supplies were donated to feed the estimated 6 million affected people. Climate alone was not responsible for the fate of these Sudanese refugees, many of whom were driven from their homes by the continuing war by southern Christian rebels against the Islamic North. (Gamma)

Right and below Drought and famine threatened over 35 million people in Africa in the mid-1980s, especially in the Sahel. In countries such as Mali, Niger and Chad, livestock numbers dropped between 40% and 90% as grasslands disappeared and streams and wells dried up. (Gamma)

places by as much as 100 km (60 miles) in a quarter of a century. Despite international efforts, by organizations such as Oxfam and Cafod, to relieve suffering in the country, the consequences of disease and malnutrition—in particular upon children and the elderly— were appalling.

After a completely dry period the first rains for many months fell in Mali in September 1974, but rainfall totals did not begin to reach 100 mm (4 in) a year again until 1976. Mali was blessed with a brief respite from 1976 to 1978, but then drought returned, becoming more severe through the 1980s. In normal times, Malian agriculture employs nearly 85 per cent of the country's population. In the early 1980s agricultural production declined drastically, mainly owing to the lack of water. Cassava, groundnuts, rice and millet shrivelled in the parched earth. The country was once again forced to live on international aid.

The rains returned to Mali in 1985, but the first good crops for years were ravaged by locusts. Again the drought returned. In 1986 and 1987 exceptionally dry conditions prevailed, although the drought was not so severe as in the years in which virtually no rain fell on Mali—1968–74 and 1976–84.

In 1988 the southern part of the Sahel enjoyed abundant rainfall and in places excessive rainfall caused some problems. For the first time in many years much of the region required little to no food aid. The northern Sahel remained dry though. The greater part of Mali is in the northern section of the Sahel, so the country was not so fortunate as some of its neighbours in that year of relative plenty.

Since then, the southern Sahel has had at least some rain each year, although the totals have been very small and there has only been one wet season since 1988. The northern Sahel is still in the grip of prolonged drought.

This change, or variation, in climate has had a profound effect upon what was already one of the poorest nations in the world. Agriculture in Mali has been decimated. Vast areas of grassland have been lost to dusty desert, stirred only by whirlwinds when the hot Harmattan winds blow out of the Sahara towards the West African coast.

Further east in the Sahel, Lake Chad is slowly drying, slowly shrinking as drought changes the face of much of the continent.

CAPE VERDE

Just over 620 km (385 miles) off the West coast of Africa are 10 rugged volcanic islands whose drought has, since 1968, become alarmingly prolonged. Cape Verde, a former Portuguese colony, is a small country, comparable in size with the English county of Suffolk and the American state of Rhode Island. It is a country whose economy and social fabric are threatened by climatic change, or by what Cape Verdeans hope is merely a temporary climatic variation.

Cape Verde enjoys moderate temperatures with an average of 22°C (71.6°F) but extreme aridity has been a problem ever since the islands were first inhabited in the middle of the 15th century. Cape Verde receives very little rain—about 60 mm (2.5 in) was the average precipitation total. This is the result of predominating Northeasterly dry winds blowing from the Sahara Desert for most of the year. Only in August, September and October does the wind occasionally arrive from a different quarter, bringing moister air and a little rainfall. For the rest of the year, the Sun may be obscured by a thick 'cloud' of fine sand blown from the Sahara.

The Cape Verdeans had learned to live with these deprivations, developing a subsistence agriculture that relied, in part, upon sea mist which provides some moisture in the hills. Large underground reserves of water were tapped to support farming and many small dams were constructed to retain as much of the meagre precipitation as possible. Under irrigation, maize, beans, sweet potatoes, sugar cane, cassava and bananas could be grown, and over three quarters of the population depended on agriculture for their living.

Since 1968, Cape Verde has experienced hot dry winds from the Northeast all year round, virtually without interruption. As a result the country has suffered almost continuous drought, broken only by torrential rain in September 1984 that perversely washed away the majority of the irrigation dams. In most years since, rainfall has either been absent or negligible, and, as a result, over 90 per cent of the country's food crops now have to be imported.

The Cape Verdean government is trying to come to terms with this severe climatic variation. With substantial overseas aid and invest-

ment, Cape Verde has embarked upon schemes in soil and water conservation in an ambitious attempt to reverse the gradual desertification of the islands and even to make them self-sufficient in food.

However, if the loss of even the scant annual rainfall of Cape Verde were to become a long-term feature, it is doubtful whether careful farming practices and additional irrigation dams could save a way of life. Cape Verde is already a land of mass exodus. About 350 000 people live on the islands, while some 100 000 Cape Verdeans live abroad—the majority in the USA and Portugal—and the money they send home is one of the country's most important sources of foreign currency.

It is too soon to know whether the almost total drought afflicting Cape Verde is a short-term climatic variation or a long-term climatic change. The islands' recent experiences have been cited in some quarters as an example of climatic change owing to global warming and, if this is the case, the exodus of Cape Verdeans may be expected to continue as their home country becomes increasingly arid.

Right and above Only careful conservation of water behind small retaining dams and the use of underground reserves permits agriculture in the chronic drought conditions of Cape Verde. Bananas, peanuts, coffee, castor beans and maize are all successfully cultivated. (Photo News)

ETHIOPIA

In many parts of the world, the climate appears to be changing, but opinions are divided about how far this change is natural and how far it is the result of human activities. In the 1980s television images of emaciated children brought a flood of relief to Ethiopia, but when the cameras had been packed up and taken away, the problem remained. The famine persisted. In 1990 most of the northern provinces of Tigre and Eritrea were subject to drought and famine.

The problem is partly man-made. Thirty years of civil war in Ethiopia have destroyed much of the agricultural infrastructure. Land has been fought over, farmers have been conscripted, seed and equipment have been unobtainable and irrigation channels have been damaged. Deforestation, too, has had an effect as the soil has been more easily eroded after the removal of the tree cover. It was estimated before World War Two that over 40 per cent of Ethiopia was forested; now under 4 per cent of the country is covered by trees.

Human activity alone is not responsible for the degradation. Climatic variation has played an important role too, and Ethiopia has an especially variable climate. Hot periods of extended drought, during which strong winds remove the dusty topsoil, alternate with cooler periods of heavy rainfall. Each period appears to last several years, although the cycle is irregular. It has been suggested that, on top of this natural cycle, there may be a long-term shift to a significantly drier climate. Figures for the country's average rainfall seem to support this hypothesis—in 1937 and 1938 Tigre received an average of 950 mm (38 in) of rain per annum; by 1973–74 that average had fallen to 475 mm (19 in). More recent, but possibly unreliable, figures indicate an even lower rainfall. These figures should, however, be treated with some caution as it is possible that the 1937–38 totals were recorded during the wettest part of a cyclic pattern.

In the areas described above it seems as though the climate is becoming drier. Near-desert conditions are taking over. In the Middle East, South America and the Indian sub-continent, too, 'desertification' is proceeding, but in each case it appears that the changes are, at least in part, due to the activities of man.

Because of the unusual plants and animals that evolved to withstand the harshness of its climate, Death Valley, California—the hottest and driest part of the USA—has been made an American national

16

monument. Occasional heavy rain can transform the desert, bringing a 'flashflood' during which long dormant seeds germinate, plants grow, flower and die all within a few weeks. (Ace and Explorer)

DESERTIFICATION

When in 1949 the French explorer André Aubreville was in the Sahel of Africa, he discovered that the savanna grasslands and the tropical rain forest were being damaged by farming activities. The land was deteriorating, the trees were being cleared and the desert was advancing. He coined the word 'desertification' to describe what was happening.

There is an inbuilt assumption in the very word 'desert' that it has changed, that it was once a better environment. The Latin *desertus* means 'abandoned' which implies that it was formerly inhabited and adequately watered for some agricultural or other activity. This, in turn, implies that deserts are man-made and can somehow be reclaimed by planting in wadis and around the fringes.

Certain plants will flourish in rainy periods in such locations but in times of drought—the usual condition in deserts (see p. 92)—they will perish. Sometimes totally inappropriate plants are grown in wadis, for example wheat which is cultivated in 'gardens' in oases in the southern parts of the Sahara. Such a demanding crop quickly uses up the supplies of underground water and as the water is salty, the salinity soon kills many of the plants. Underlying theories that plants can hold back or reclaim the desert is a misconception about the very nature of the desert and the climatic, or occasionally man-made, change that created it. Deserts are a natural feature. The desert landscape, its soils and what little flora there is are a perfect adaptation to that climate. By assuming that the deserts are only man-made, rather than being a response to climate, we are in danger of concluding that it is possible to reclaim deserts on a large scale, although surface features such as vegetation can have some effect on atmospheric circulation.

In marginal areas such a reclamation may be possible. It all depends upon the aridity index of the area. All deserts are dry, but some are drier than others, that is they have a greater aridity index. Death Valley in California is often regarded as the archetypal desert, especially in the USA. This impression is added to by the fact that it is the hottest place in the world with an American maximum temperature of over 48.9°C (120°F) being recorded on 43

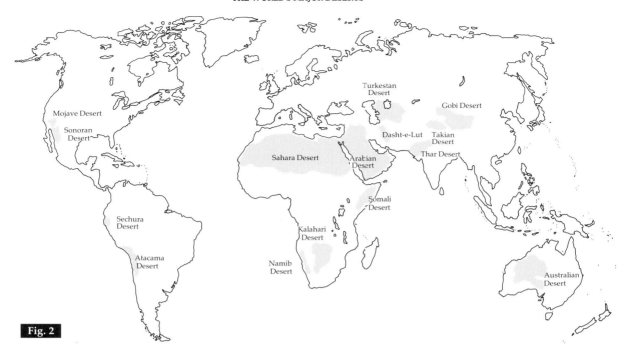

Fig. 2

WHAT MAKES A DESERT?

A desert is an area in which the land surface is dry and in which there is little to no vegetation owing to the deficiency of water. A lack of rainfall is the normal cause of desert conditions and most deserts are in areas of high pressure from which air flows out and little to no moist air flows in to bring rainfall. Some deserts are caused by continentality, that is they are so far from the sea that the air masses that reach them have lost their moisture already. Many deserts—such as the Sahara and the Arabian Desert—are hot, and this increases the lack of water owing to the high rates of evaporation. But it is the lack of water that makes a desert, not temperature, and the icy wastes of the Arctic and the Antarctic are often referred to as polar deserts because there is no surface water there in a liquid form. Yet other deserts are caused by being in a rain-shadow—the Patagonian Desert, for example, is dry because it lies at the foot of the Andes mountains over which the prevailing winds have already passed, losing most of their moisture as rainfall and snow on their passage over the mountains.

consecutive days between 6 July and 17 August 1917. Yet Death Valley experiences occasional heavy storms which bring large amounts of precipitation in a few hours. It has an aridity index of seven, which means that the Sun would evaporate seven times the average rainfall it actually receives. The Sahara Desert is much drier—the driest area on Earth. In parts of Algeria, Libya and Niger the aridity index is 200, meaning that the Sun would be capable of evaporating 200 times the rainfall received. Thus, all deserts are different. Some are not so dry and might be easier to reclaim or perhaps we should say 'claim'. This does not mean, though, that we can have any great confidence in being able to undo the work of desertification even if human activity is the principal agent. Yet if a climatic variation is the principal cause, and observations in the western Sahel suggest that a natural climatic variation is the cause, it is not inconceivable that, with very careful husbandry and conservation, some areas could be reclaimed if a wetter period arrives. The situation in, for example, Ethiopia is somewhat different as degradation has been, in part, the response to human activity such as the clearing of forests.

INLAND SEA

In March and April in 1988, 1989 and 1990 abnormally heavy rain fell on the eastern side of Australia. The bewildering pattern of freak weather continued in April 1990 when large areas of eastern Australia suffered severe flooding following several weeks of record-breaking torrential rain. The districts affected—in inland Queensland and New South Wales—normally receive only light to moderate rainfall, but in less than one month some places were deluged with their usual annual rainfall total. Ironically, the area had been attempting to cope with a prolonged drought—no significant rain having fallen since July 1989.

About 200 000 sq km (77 000 sq miles) were inundated, creating what could be mistaken for an 'inland sea', as swollen rivers broke their banks and flooded an area as big as Britain. Four people drowned, many sheep were lost and dozens of small country towns were cut off by the rising water. The 2000 inhabitants of Nyngan, on the Bogan River, in New South Wales were evacuated in the biggest airlift in Australian history. Townspeople had unsuccessfully fought to hold the flood back with sandbags.

The Australian air force flew in emergency supplies to the Jericho district, 800 km (500 miles) Northwest of Brisbane, and dropped fodder to stranded sheep and cattle around Charleville and Cunnamulla in Queensland and Walgett in New South Wales. Rail services throughout inland eastern Australia came to a standstill and hundreds of miles of road were flooded. Minor flooding affected Sydney and Newcastle, and for the third year running the cotton crop of the Namoi area in New South Wales was spoilt by unseasonal heavy rain.

The Australian Bureau of Meteorology acknowledges that a pattern is emerging but stresses that it is too early to be talking about a climatic change. What is the cause of the freak weather? For once, global warming is not thought to be the culprit. An unusually strong ocean current brought warm tropical waters much further South down the East Coast of Australia during the first quarter of the three years in question and it seems probable that the increase in sea temperature off Australia was the reason for the change in the pattern and character of air masses that brought the heavy rain to the country. The current is thought to be connected with El Niño, the mysterious Pacific Ocean current which is examined in *Current research* (see p. 164).

NO SNOW

The winter of 1989–90 was the third in succession during which the Alps received a greatly diminished snowfall than normal. Skiers in Europe desperately sought the few lengthy pistes in the minority of Alpine resorts high enough to guarantee constant snow. Disappointed tourists, lured to the mountains by brochures filled with pictures of long slopes of crisp snow, were forced to take up summer sports such as canoeing and hiking.

French skiing resorts were particularly badly hit. Some Alpine and Pyrenean villages were reported to be on the verge of economic ruin if the mild weather that resulted in snowless slopes continued. Over the

Nyngan, a small town of under 3000 people in New South Wales, is situated in a semi-arid zone in which abundant rain is exceptional and dry conditions are normal. It is too far south to benefit from the bore holes of the Great Artesian Basin and is used to chronic drought. The irony of the floods of April 1990 cannot have been lost upon its inhabitants. (Gamma)

Left and below Skiing—a method of transport in Norway and Sweden for thousands of years—did not become a sport, and the economic mainstay of dozens of Alpine communities, until the 19th century. But just as rising standards of living, increased leisure time and the growth of the package tourist industry brought prosperity to places such as the French Alps, the failure of the snows in 1988, 1989 and 1990 brought unemployment, bankruptcy and questions concerning climatic change. Only the highest resorts and those that had invested in snow cannons (below) prospered. (Gamma)

Christmas and New Year period, in the Pyrenees only 20 per cent of French resorts had enough snow to be able to open their pistes. In France as a whole less than half the country's ski centres bothered to open during what should have been their busiest period of the year, and only 20 French ski resorts were high enough to be functioning normally.

The SNTT (Syndicat National des Téléphériques et Téléskis)—the union that represents French workers employed on cable lifts—requested government assistance, asking for urgent financial help for the snowless villages. Less than a quarter of the ski lifts were functioning in French resorts during the 1990 New Year holiday and some 10 000 seasonal workers in the skiing tourist industry were laid off.

A number of resorts, including Chamonix, had taken precautions to protect their winter tourist trade following a couple of unusually mild winters. Chamonix itself was too low to have any snow on its slopes, but in 1985 a local businessman had invested in 33 snow cannons which were brought into use to create a 300-metre (1000 ft) strip of artificial snow on the town's nursery slopes. Although half of Chamonix's high level pistes were open, the arduous trek to find snow proved too much for many holiday-makers who enjoyed the clear blue skies and the attractions of the tennis courts and golf club instead of more normal winter sports.

The lack of snow spelt trouble for Austrian and Swiss resorts too. Most ski centres in Austria are not at any great altitude and for part of the 1989–90 season, few had snow in sufficient quantity. Only centres near glaciers, such as the Stubai and Solden near Innsbruck, the Arlberg region and parts of Salzburg, were unaffected by the mild weather because of altitude. When snow did come to the Austrian Alps in the first week of January, it was too late to prevent considerable losses of income to hoteliers, ski instructors and cable car operators. Tourist boards and skiers agreed that it was the worst winter sports season for 30 years.

The best pistes were in Italy, not because temperatures were any lower or precipitation any heavier in the Italian Alps but because so many Italian resorts had invested in expensive snow cannons to provide an artificial covering.

Higher resorts, especially those in the Val d'Aosta, were operating normally with long pistes and a snow depth of between 25 and 35 cm (10 and 14 in), but at Monte Livata, the Apennine ski resort East of Rome, horse riding was more in evidence than skiing.

By the middle of January some snow fell in the French Alps and Pyrenees, enabling more resorts to open, but this late fall initially provided only poor, light cover because there was no base of powdery snow underneath to hold it. By the end of the month the weather had changed completely—temperatures had dropped sharply and the ski resorts of Europe had adequate snow on their slopes, and heavy falls later in the season brought avalanches.

If much of Europe was short of snow during the winter of 1989–90, it was a lack of rainfall that caused problems during the following summer. Most of the continent was hit by a prolonged drought.

DROUGHT IN EUROPE

The parched conditions during the summer of 1990 were cited by some observers to be further heralds of climatic change. On Friday 3 August a temperature of 37.1°C (98.8°F) was recorded in Cheltenham, Gloucestershire in the West of England, and several amateur meteorologists reported temperatures in excess of 100°F (37.8°C). Big Ben stopped owing to the heat and the surface of several British motorways melted. Bewildered by the searing heat, Britain hardly noticed the extraordinary drought that held a desiccating grip over much of Europe.

In Greece the water shortage was the worst for over 50 years. In order to conserve dwindling supplies, the Greek government trebled the price of water to consumers and introduced fines equivalent to £4000 for washing cars or using hosepipes on gardens. The effect upon agriculture was severe. The production

Right The drought of 1990 continued into the late autumn. Water rationing affected several major Italian cities and across southern Europe many rivers were reduced to a trickle. In the French département of Aveyron, for example, rivers became rivulets and hydro electric power plants on rivers such as the Truyère were unable to function at capacity. (Gamma)

of most crops was halved and tomatoes, one of Greece's most important exports, had to be imported. Parts of the Peloponnese resembled a dusty desert and many farmers were brought close to ruin. The Greek premier Konstantinos Mitsotakis described the extreme drought as a 'national disaster'.

In France temperatures of 40°C (104°F) were recorded near Toulouse. The cities of southern France blistered in a suffocating heat and a number of elderly people died of the effects of sunstroke. French newspapers sported photographs of the camels introduced into one village by a landowner who had experienced drought for nearly half a dozen years, while other frustrated farmers sabotaged water pumping stations, anxious to divert domestic supplies to irrigate their holdings. The level of rivers fell all over the country. The Garonne was reduced to less than half its normal flow and its remaining waters were so warm that they could not be used to cool the Goflech nuclear plant.

Most of Italy received little to no rain between May and late August and this added to the problems caused by a dry winter. In the worst drought for 250 years, precipitation totals fell by nearly 50 per cent. In the Mezzogiorno (southern Italy) and parts of central Italy crops shrivelled and died in the fields. So little rain fell in the region of Campania that in July domestic water in Naples was rationed to a couple of hours a day, while in Sicily the price of bottled water exceeded that of beer. Rome endured high temperatures for weeks, and the drought extended as far North as the rich farming regions of Lombardy and Emilia Romagna for the first time in the 20th century. The weather situation in Italy was confusing though. Central Italy received rainfall totals only slightly below normal, but the bulk of this precipitation came at the wrong time of the year for it to be of any use for crops.

Southern Europe and North Africa suffered its third year of prolonged drought, and in Algeria and Tunisia the water shortages reached the proportion of national catastrophes until relief came with heavy rainfall associated with thunderstorms at the beginning of August. Spain, however, avoided the worst of the ordeal. Although the fields of Andalusia and Valencia were as dry as most of the rest of southern Europe, rainfall did come at the crucial time for Spanish agriculture. Even Almeria, which is usually the driest region of Spain, received a 50 per cent increase in precipitation, although its mean rainfall between June and August is normally so low that an increase of even 50 per cent was not significant.

Other parts of the world were also experiencing abnormal weather conditions. At the end of July parts of New South Wales received rainfall continuously for almost two days.

Nowra Air Station, near Sydney, recorded almost 100 mm (4 in) in 48 hours. Coastal regions of New South Wales suffered widespread flooding.

The northern edges of the Sahara normally receive about 100 mm (4 in) of rain a year. At the beginning of August intense thunderstorms brought the usual annual rainfall total in a couple of days, and Touggourt, in Algeria, recorded nearly 115 mm (4.5 in) in 72 hours. In the same period no less than four typhoons were active in the northern half of the Pacific Ocean. No one seemed to notice as none of the hurricanes touched land, but the meteorologists recorded their progress by satellite and noted yet another variation to what was regarded as the climatic norms.

The reason for the high temperatures and extraordinary drought in western Europe was the lingering presence of large anticyclones, or highs, which covered the greater part of the western and southern parts of the continent during the first half of 1990. These areas of high pressure disappointed skiers with the consequent lack of snow during the Alpine winter at the beginning of the year. They brought clear skies and killing frosts to much of Europe in the spring, and then gave the continent the extended drought of the summer.

Western Europe's normal weather pattern is one of continuous change. Fronts and lows constantly move in to Europe from off the Atlantic, regularly interrupted by areas of high pressure (anti-cyclones). Each brings their own particular characteristics.

A depression, an area of pressure lower than the surroundings, occurs in a region where air of different temperatures is in conflict. Depressions—popularly known as lows but also called cyclones—bring predictable weather as each sector of the depression passes overhead.

Fronts form along the boundaries of the cold and warm air. Uplift of air—and cloud formation—occurs along these fronts and other parts of the cyclone as the warm air rises. Air is drawn towards the centre of the depression in order to fill it, and these movements can be seen on a weather map in the flow of wind around the low. From satellite photographs a low over western Europe can normally be identified as a giant anti-clockwise swirl of clouds. (In the southern hemisphere, the movement of air around a depression is clockwise.)

Ahead of the warm front, as a low passes over to the North of Britain, the wind blows from a general southerly direction. The strength of the winds depends upon the pressure gradient, which is the rate at which pressure descends towards the centre of the area of low pressure. (The wind speed can also depend upon the intensity of a high.) This can be identified upon a weather map by the isobars—lines which join points experiencing the same level of atmospheric pressure. If the isobars on the weather map are close together the pressure gradient is steep and the winds will be strong as the air moves rapidly to fill the low. If the isobars are further apart, the pressure gradient is less and the winds will be lighter.

As the warm front of the depression approaches, the sky becomes cloudier. First high wispy cirrus and cirrostratus clouds will usually appear; then, lower darker altostratus and finally dark, dense nimbostratus clouds bring rain as the warm front passes.

Behind the warm front the skies clear in the warm sector of the depression with only patchy stratocumulus cloud, although over hills heavy rain may continue. Temperatures rise and the predominating wind blows from the Southwest, its strength depending upon the pressure gradient. As the cold front approaches, the cloud cover becomes thicker. Immediately ahead of the front towering anvil-shaped cumulonimbus clouds bring heavy rainfall.

Behind the cold front the temperature drops suddenly as cold air predominates. The sky becomes clearer with only intermittent fluffy white cumulus clouds, and the wind changes direction again with northerlies predominating. This is the 'ideal' pattern or classic model for weather conditions on the passage of a cyclone. But most cyclones differ in detail and the track followed by the cyclone is important.

This regular passage of lows, interrupted by the occasional dominance of an area of high pressure (or anti-cyclone) gives western Europe its varied weather. An anti-cyclone is an area in which air is sinking rather than rising. In such an area, therefore, water vapour is not carried high enough for deep clouds to form and, thus, highs are, in general, dry. Much

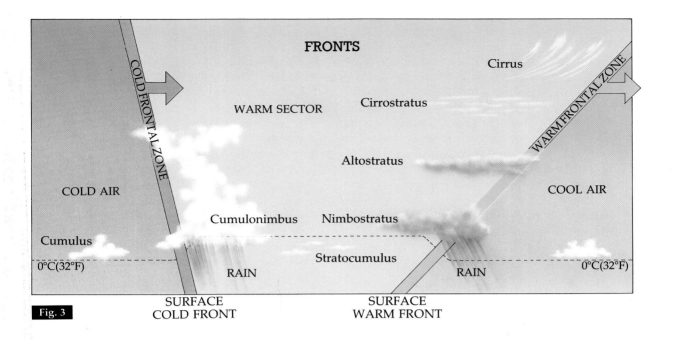

FRONTS

Cirrus

COLD FRONTAL ZONE

WARM SECTOR

Cirrostratus

WARM FRONTAL ZONE

Altostratus

COLD AIR

COOL AIR

Cumulonimbus   Nimbostratus

Cumulus

0°C(32°F)

Stratocumulus

RAIN   RAIN   0°C(32°F)

SURFACE   SURFACE
COLD FRONT   WARM FRONT

Fig. 3

will depend upon where the high is centred. Highs over the Bay of Biscay, for example, give mild weather in the winter although usually in the winter, highs bring cold weather to western Europe. In the summer—as in the summer of 1990 when a large enduring anti-cyclone dominated much of the continent for months—highs result in hot and sunny weather. Winds blow out from these often long-lasting anti-cyclones in a clockwise direction. (In the southern hemisphere, the wind blows anti-clockwise in an anti-cyclone.)

For much of the first half of 1990, the normal passage of air masses over the greater part of Europe was disrupted by a 'blocking anti-cyclone'. Blocking has been under investigation by meteorologists for over 40 years, but the extent of the phenomenon and its effects are only now being realized. It used to be thought that blocking was confined to the temperate latitudes—such as western Europe—in which the usual flow of westerly air masses were blocked by a large and enduring anti-cyclone. It is now known that blocking is experienced over a wider area of temperate latitudes and that its effects have worldwide implications on weather.

The predominating westerly winds over Europe fluctuate in strength between powerful, nearly constant flows and those following a weaker more erratic path. Given such a degree of instability, the upper levels of these westerlies may bifurcate, trapping an area, or block, of high pressure between the two branches. The static blocking anti-cyclone—which results in remarkably stable weather conditions within the area it covers—may last for as little as a couple of weeks, but in the first half of 1990 they endured for months at a time over western Europe. Over the short term, blocking anti-cyclones are maintained by the momentum of the lows passing on either side. It is not, however, yet known with any degree of certainty how blocking anti-cyclones are able to last for as long as several months, although it does seem that the speed of the winds at higher levels may be an influencing factor.

Climatic research has correlated the incidence of blocking anti-cyclones and a number of unusual climatic variations including the severe winter of 1977 in parts of the USA and Canada when a prolonged anti-cyclone brought extremely cold conditions over a considerable spell. And blocking, rather than global warming, appears as the immediate cause of the extraordinary conditions experienced in Europe in 1990. Indeed, no link between the incidence of blocking anti-cyclones and global warming has been shown. It is, however, improbable that blocking occurs in isolation of other phenomena.

RETURN TO NORMAL

The drought in Europe continued through most of the autumn of 1990. As late as November, pictures appeared in the press of pools where there should have been reservoirs. Italy was especially badly hit and some large Italian cities were effectively under water rationing almost until the end of the year.

There seemed, however, to be no end to the freak conditions. In Southern England during the autumn some households suffered the indignity of flooding after heavy rain, but their local drought orders remained in effect.

Yet by the end of the year a return to normal seemed to be in the air. By the last week in November there had been heavy snowfall in many parts of the Alps. A few passes were already blocked and hoteliers were looking forward to a good skiing season. Further substantial snow was forecast and if there was another example of freak weather, it indicated colder, not warmer, conditions.

In November 1990 Lugano in Switzerland experienced most unusual weather. Sheltered between the 'sugar loaf' peaks of Monte San Salvatore and Monte Bre, the Swiss resort of Lugano enjoys an almost Mediterranean climate. The gardens that line the promenade along Lake Lugano boast palm trees, oleanders, canna and other warmth-loving plants. Throughout the winter there is a mildness in the climate that has helped the town to flourish as a resort. Yet in November 1990 Lugano was blanketed in snow—not a mere dust-like covering but rather a substantial thickness. No one could remember anything like it, and once again the question was posed 'What is happening to the weather?'

By 1991, North America was returning to normal, following a major drought in the 1980s. During 1988 much of the Mid-West experienced a drought almost as intense as that suffered in the 1930s when the 'Dust Bowl' was produced. The harvest of 1988 in the Mid-West was over 30% down on that of 1987. Farmers went bankrupt in great numbers, making the drought the costliest natural phenomenon in American history.

IS SOMETHING STRANGE HAPPENING TO OUR WEATHER?

The accounts that you have read in this chapter would seem to suggest that a pattern of freak weather is building up—a pattern that could be the first evidence of longer term climatic

1990—A YEAR OF FREAK WEATHER

January	Mild weather throughout Northern Europe. Alpine ski slopes snow-free. Fierce storms over Britain and NW Europe.
February	Unusually mild weather in Scandinavia—Stockholm harbour ice-free. Snow in Florida. Fierce storms over Britain and NW Europe.
March	Drought in much of Europe.
April	Torrential rain floods semi-arid region of Australia.
May	An unusually long, hot summer begins in most of Europe.
June	Drought in Southern Europe intensifies. Severe drought in Greece.
July	Record high temperature in France.
August	Record high temperatures in England. Severe drought in Algeria and Tunisia.
September	First substantial rainfall in much of Europe since March.
October	Fierce storms in the North Atlantic.
November	Heavy snow in normally snow-free areas of Southern Switzerland.
December	Record low temperatures in India. Severe storms in Australia.

change. In the next two chapters are descriptions of the unusual weather conditions that have been experienced in Britain and Northwest Europe in recent years: unusually long hot summers and months of drought, and sudden storms of quite unusual ferocity during the winter.

In later chapters possible explanations are examined, most notably global warming—the so-called 'greenhouse effect'. In many quarters, greenhouse or global warming is accepted as a fact—others remain somewhat sceptical. The facts and the opinions are recorded in this book, but it is up to you, the reader, to come to your own conclusions from the evidence. Perhaps it is that, in the view of many climatic researchers, it is too soon to tell. Others, however, warn of catastrophe.

At this point, it may be worth asking whether we are not subject to a fashionable myth, for there are observers who tell us that we have been here before. On 3 June 1975, in a leader entitled 'The Prospect of a Minor Ice Age', *The Times* declared:

'If the cooling trend continues through the spring and summer seasons, crops will be reduced and the possibility eliminated of introducing varieties of maize, soya and other breeds into what have been for them marginal climatic regions like the South of England and North of France.'

For much of the 1970s, there was grave concern about the possibility of global cooling, and one of the cult environmental books of the decade was *The Cooling*.

It is strange that so many serious predictions of global cooling were being made in years that enjoyed notably long warm summers—1975 and 1976. But the general trend in the 1970s was towards cooler conditions and based upon the conditions that could be observed at that particular time, long-term forecasts of a 'minor Ice Age' were made.

The conditions described in this chapter are of drought, of milder winters, of high summer temperatures and of freak isolated climatic events, such as torrential rain causing floods and intense storms. These are the conditions being observed now, and it is partly upon them that current long-term projections are being made. It is worth reflecting whether these forecasts will prove to be any more accurate

than the reverse scenario envisaged in the 1970s.

It is revealing, too, that the immediate climatic conditions since the mid-1970s have led to the formulation of theories that end in an imminent catastrophe—although the predicted outcome is very different, it is nevertheless catastrophic and one begins to wonder if there is not a basic human delight in predicting doomsday. There have always been those who have predicted disaster, whether they be millenarians or scientists. The 18th-century clergyman Thomas Malthus forecast economic catastrophe, after he had calculated that while the number of people on Earth increases geometrically, the amount of food that they can produce only increases arithmetically, that is at a slower rate. He predicted famine, war and disease. What he did not—and could not—foretell were the technological and other innovations that would increase the quantities of food that could be grown, and the medical and social advances that—in the developed world at least—would slow the rate at which the population increased. But, at the time, Malthusian doom was widely accepted as truth.

A similar theory of disaster stalked the second half of the 19th century. North America and Europe were growing rich as a result of the Industrial Revolution whose power base was coal. The economist Jevons predicted catastrophe when the coal ran out—which in the 1870s seemed likely. He cannot have suspected the later discovery of more and more coal, nor could he have been expected to predict the oil economy, the advent of hydroelectric power or nuclear power.

In the 1960s the first fashionable scare was once more the population explosion. Today disaster is predicted because of man's economic activities, because of the destruction of the forests, the pollution of the air and the emission of greenhouse gases. The theories about what is happening and what might happen to our changing world of weather are highlighted in this book, but it is worth remembering that we are basing our theories and our understanding upon current technology. What we cannot know are what advances might be made in the next 50 years that could change the situation out of all recognition—in *either* direction.

2 The Tempest

The night of 15–16 October 1987 was the most traumatic in the lives of many people living in the Southeast of England. It was the night that alerted them to the possibility of climatic change because the gusts of wind that tore the heart out of the Home Counties in the early hours of Friday 16 October were so abnormal, so far outside the experience of the majority of the population.

The sky was overcast that Thursday evening. There seemed every prospect of rain, after all there had been a heavy downpour for most of the day. It was windy. Everyone assumed that a gale was brewing, but the immediate worry for people living in low-lying areas was flooding, and in some districts a flood alert was issued. No one suspected what was to come.

As the evening wore on, the pressure fell in a deep depression over the Bay of Biscay. As the low came North it intensified dangerously and at midnight the fierce winds that tore round the South and the East of the low reached the coast, hitting the Channel Islands, the Isle of Wight and Portsmouth. The results were devastating. It was, quite simply, the worst gale in living memory and probably since The Great Storm of 1703.

The centre of Brighton resembled the after effects of a bombing raid with roads blocked, heavy structural damage and virtually all the fine trees in the town centre felled by the wind. The hangar at Southend Airport collapsed, crushing the aircraft trapped inside. A caravan site at Peacehaven, high on the cliffs in East Sussex, was wrecked as 200 mobile homes were lifted and shattered as easily as if they had been matchboxes. Throughout Southeast England the pattern was repeated.

Sussex was the worst hit county. Large areas of the Weald were cut off by fallen trees; roads and railway lines were impassable; telephone lines and electricity pylons and cables were down. Communications were disrupted for days and some villages were without power for a week. A 'wartime' camaraderie helped to raise the spirits in some villages deprived of power. People gathered in parish halls and community centres to cook, sometimes for days after the hurricane. Roofs were torn off, chimneys toppled, fences flattened, windows broken. Church towers and spires, for example at Rotherfield, came down, and Newhaven and Lancing were only two of the places in Sussex at which homes were virtually destroyed.

Over 15 million trees were lost in Southeast England that night. In one wood near Midhurst, in West Sussex, only half a dozen trees remained standing, while at Toys Hill, near Westerham in Kent, 99 per cent of the trees were blown down. Chanctonbury Ring, a circle of trees perched up on the crest of the South Downs and to many people the symbol of Sussex, was ruined—so many trees were lost that the familiar outline of the Ring became just a ragged clump overnight. At Sevenoaks in Kent, the famous seven oaks at the Vine cricket ground were reduced to one. Whole copses were toppled with large and ancient trees felled like skittles, their gnarled roots upturned and exposed to the air.

The grounds of stately homes and gardens were wrecked. Kew Gardens lost many irreplaceable specimen trees. The famous gardens at Sheffield Park in East Sussex lost over 2000 trees. In many cases the damage to the woodlands had not been repaired three years later and some footpaths were still blocked. The face of the countryside of Sussex, Kent and Surrey was changed by The Great Storm. New views were opened up overnight as woodlands disappeared and the memory of how things were before the hurricane is precious to many country people.

The wind speed reached 167 km/h (104 mph) at Herstmonceux in East Sussex, at Dover, Kent and St Catherine's on the Isle of Wight. At Shoreham in West Sussex a gust of 174 km/h (108 mph) was unofficially recorded. The height of the storm over the South Coast, the chalk Downs and the Weald was between two and three o'clock on the Friday morning. The wind rattled like thunder as the deep low moved away to the North and East bringing the gale to London, and then in a weakened form—but with winds still gusting over 113 km/h (70 mph)—to East Anglia and the East Midlands.

The most ferocious blasts came in sudden gusts lasting less than 30 seconds, with calmer periods of 'normal' gale force winds between. The strongest winds were to the South and East of the deep depression, where the isobars were closest together (see weather map on p. 34).

The coast of Sussex was badly hit by The Great Storm. At Littlehampton an entire road of houses was made uninhabitable, storm damage which made the destruction of beach huts appear trivial. (Portsmouth Pub. & Printing Ltd)

THE BEAUFORT SCALE FOR MEASURING WIND STRENGTH

Wind force (the Beaufort numbers)	Description	Wind speed km/h	mph
0	Calm	less than 1	
1	Light air	1– 5	1– 3
2	Light breeze	6– 11	4– 7
3	Gentle breeze	12– 19	8–12
4	Moderate breeze	20– 29	13–18
5	Fresh breeze	30– 39	19–24
6	Strong breeze	40– 50	25–31
7	Near gale	51– 61	32–38
8	Gale	62– 74	39–46
9	Strong gale	75– 87	47–54
10	Storm	88–101	55–63
11	Violent storm	102–117	64–73
12	Hurricane	118+	74+

The scale has unofficially been extended to Force 17 to describe tropical storms.

By the evening of 15 October 1987 western France was experiencing high winds and the word 'hurricane' had been used on the French weather forecast. The Great Storm hit Brittany before it reached Britain, causing almost as much damage as it did in southeast England. In countless Breton coastal villages houses were damaged, trees felled and boats tossed like driftwood. (Gamma)

WEATHER MAPS

The weather map opposite is an extreme example. It portrays the weather at a particular moment, in this case during The Great Storm of October 1987.

Weather maps are more correctly known as pressure charts. The lines that cover the map record pressure and are called isobars. They join points experiencing the same atmospheric pressure. The pressures shown are recorded in millibars (mb)—1 millibar equals 100 newtons per square metre.

The number shown at the centre of the depression is a reading of the low pressure. Areas of low pressure are usually indicated by the letter L on a weather map; areas of high pressure are marked with the letter H.

The closer together the isobars, the more strongly the wind is blowing. Isobars close together indicate a steep pressure gradient. The wind blows towards the centre of the low—in the example given here particularly fiercely. Over the sea the wind flows almost parallel to the isobars; over the land the wind flows at more of an angle towards the isobars, as can be seen on the map from the symbols indicating the wind strength and direction. The 'wind

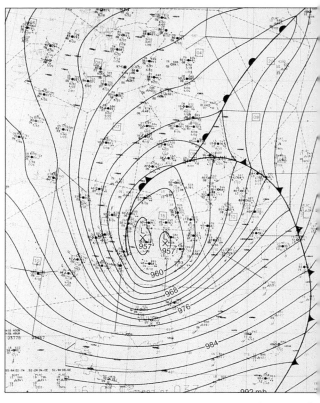

The pressure map at 0300, 16 October 1987. Southeast England is receiving the brunt of the ferocious winds (above). In the second map showing the situation at 0600 the centre of the deep depression has moved on and the storm has hit East Anglia and Belgium. (Crown copyright)

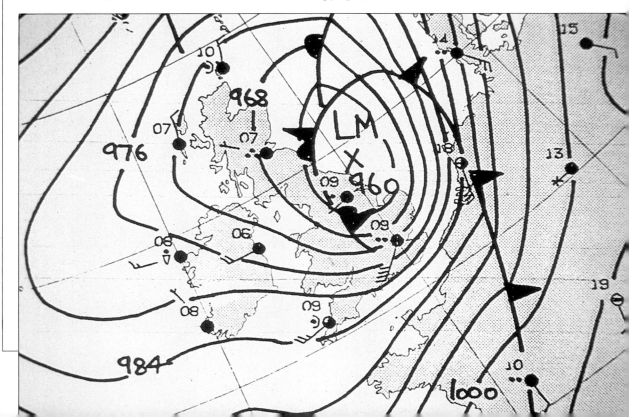

arrows' show the direction, with the end of the arrow without markings indicating the direction towards which it is blowing. The strength is shown by a series of marks on the tail of the arrow:

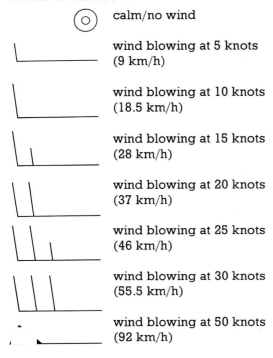

calm/no wind

wind blowing at 5 knots (9 km/h)

wind blowing at 10 knots (18.5 km/h)

wind blowing at 15 knots (28 km/h)

wind blowing at 20 knots (37 km/h)

wind blowing at 25 knots (46 km/h)

wind blowing at 30 knots (55.5 km/h)

wind blowing at 50 knots (92 km/h)

The data shown on the map is recorded at weather stations some of which are indicated by a circle. Shading within the circle shows the amount of cloud cover experienced at that weather station.

Cloud cover is measured in oktas, literally eighths and is expressed in terms of the number of eighths of the sky covered or obscured by clouds.

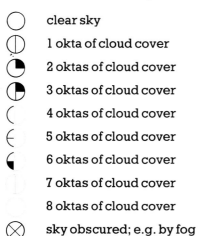

clear sky

1 okta of cloud cover

2 oktas of cloud cover

3 oktas of cloud cover

4 oktas of cloud cover

5 oktas of cloud cover

6 oktas of cloud cover

7 oktas of cloud cover

8 oktas of cloud cover

sky obscured; e.g. by fog

Beside the circle indicating a sample weather station, various symbols may be found. These represent the weather conditions that are being experienced. The nine common symbols are:

● rain   ⊿ ice pellets

drizzle   ⌒ dew

▽ showers   ≡ fog

snow   thunderstorm

▲ hail

The most prominent features to appear on any weather map are the fronts—cold, warm, stationary and occluded. A front is an interface between two air masses of differing temperatures. They are represented by the following symbols:

cold front—this is the interface between a cold air mass which is overtaking a warm air mass.

warm front—this is the interface between a warm air mass which is overtaking a cold air mass.

stationary front—this is the interface between two air masses of a similar temperature.

occluded front—this represents the place where a cold front has overtaken a warm front.

The symbols shown are internationally agreed and can be seen on weather maps from around the world. When plotted on weather maps, they provide essential information used in making weather forecasts. Simplified versions of these symbols can be seen on the charts used in television weather forecasts.

Left After 16 October 1987, thousands of claims for damage poured into the offices of insurance companies, as chimneys had toppled, trees fallen and walls blown apart like confetti, but one British insurance company—based far from the storm damage—was initially unsympathetic to claimants, dismissing The Great Storm as 'just a high wind'. (Gamma)

Right Toys Hill near Westerham in Kent after The Great Storm of October 1987. West Kent was particularly badly hit by the full force of the wind and the town of Sevenoaks became 'one oak' when six of the seven famous oak trees were toppled in the gale. (Aerofilms)

WAS IT A HURRICANE?

It was fortunate that The Great Storm came in the early hours of the morning when there was little traffic about and most people were at home in bed. Nineteen people were killed and the total would undoubtedly have been heavier if the gale had happened during the daytime. Southeast England was shaken in more senses than one by The Great Storm or what came to be known—however in-accurately—as 'The Hurricane'.

The London Weather Centre stoutly maintained that this, the fiercest storm to hit England for over 280 years, was a 'vigorous depression'. A hurricane, they maintained, is a tropical cyclone that occurs in the Caribbean, with characteristics other than strong winds such as an 'eye' and pressures that can fall below 900 mb. On the Beaufort Scale, a hurricane is a wind of over 118 km/h (74 mph) though. Popular acclaim and the *Oxford Dictionary* ('hurricane a wind of 73 mph or more') also disagree with the London Weather Centre. It may have been merely a vigorous depression, but nothing will ever persuade the people of Sussex that they didn't survive a hurricane.

HURRICANE PAST

The winds of October 1987 were not unprecedented in southern England. There are records of terrible storms during the Middle Ages, but the most recent comparable gale—until January and February 1990—was in November 1703.

Two weeks of severe gales heralded The Great Storm of 26 November 1703. A series of deep depressions swept up the English Channel, bringing appalling damage to the whole of southern England. The Great Storm of 1703 sunk the Channel Squadron of the Royal Navy. HMS *Mary*, *Northumberland*, *Restoration*, and *Stirling Castle* went down on the Goodwin Sands with a total loss of life of nearly 1200. Three other ships of the line were lost elsewhere with many fatalities. On land, hundreds died as trees blew down and houses collapsed. The Bishop of Bath and Wells was crushed to death when a chimney stack came through his bedroom ceiling. Queen Anne cowered in a cellar while her palace suffered considerable structural damage.

The Great Storm of 1703 came at the same time of night as The Great Storm 284 years later. The damage and loss of life in 1703 was

far more severe than in 1987, but we can only guess that the wind gusted with a similar or perhaps even greater force. Daniel Defoe chronicled the events of November 1703 with great verve, commenting that many people thought that the world was coming to an end. Their fear of Doomsday was echoed by more than a few, 284 years later.

A REPEAT PERFORMANCE

No sooner had people assured themselves that The Great Storm of October 1987 was a 'one-off', than a second storm of terrible ferocity hit Britain just over two years later, to be followed by a third within a few days.

The intense storms of early 1990 were widely reported by radio, television and the press. Predictably, reports concentrated on the experiences of the Home Counties—the closure of roads and railway lines, the tragic accidents, the structural damage caused to comfortable suburban properties.

The stronger winds that buffeted the North did not seem to receive such attention from the media, and the fiercest gales that swept across the Hebrides and Northwest coast of Scotland appear as a footnote in the story of the storms. Winds higher than those that scoured the South East during The Great Storm of October 1987 raced across the Highlands.

Late on the evening of Tuesday 30 January 1990 a gust of wind may have reached the record speed of 259 km/h (161 mph) at the Butt of Lewis in the Western Isles. That, at least, was the claim of some locals. Regrettably, the claim will remain 'not proven' as there seemed to be 'gremlins' in the windspeed recorder at the Butt of Lewis lighthouse. 'Something . . . loose' in the mechanism of the anemograph prevented an undisputed record from being confirmed, although the observations of the staff at the meteorological office at Stornoway, the capital of the Western Isles, show that a wind of very great speed was attained during the Hurricane Force 12 storm.

The end of January and the first days of February in 1990 witnessed a bout of freak weather conditions. Storms of very great intensity battered much of Britain and left a path of destruction across northern France, Belgium, the Netherlands, Luxembourg, the North German Plain and Denmark.

The storm claimed 46 lives in Britain and a further 49 in northern Europe. It was estimated that insurance claims amounting to over one billion pounds were received by insurers in Britain and on the continent.

The deep depression which brought the storms of January 1990 differed from The Great Storm of October 1987 in a number of respects. The storm was not so severe, the winds did not attain such a great speed, but there were far more casualties because the gale came during the daytime whereas the October 1987 storm arrived during the early hours of the morning. The vigorous depression at the

The Butt of Lewis—the most northerly point of Harris-Lewis, the largest island in the Outer Hebrides—has long been called Britain's windiest place. The weather station at the Butt consistently records high wind speeds, although the record wind speed in the United Kingdom was 278 km/h (172 mph) on Cairn Gorm Summit on 20 March 1986. The world record is 371 km/h (231 mph) at Mt Washington, New Hampshire, USA, on 12 April 1934.

centre of the storm was also far less deep in January 1990—dipping just below 970 millibars—than that of The Great Storm of 1987 when the pressure plummeted to under 960 millibars.

The path of the storm centre was also much further to the North, passing through Scotland, thus a much greater area of the country was affected by the winds.

HOW DID THE STORMS COME ABOUT?

The storm of January 1990, like The Great Storm of 1987, was simply the result of the normal method by which gales begin in the North Atlantic. Cold polar air came up against warm tropical air and along the front between them, the great temperature differences created a deep depression. Conditions in the upper atmosphere were also very favourable to intensify the cyclone. The vigorous low that resulted, registered wind speeds far lower than those associated with a similar depression, formed at almost the same time, that recorded winds in excess of 225 km/h (140 mph) over Scotland (see Fig. 4).

Some scientists suggest a different scenario. They state that global warming has increased the surface temperatures of the oceans in the tropics to such an extent that when water from the oceans evaporates, it rises higher than usual into the air. This creates stronger than normal air currents which are able to penetrate further North than before, bringing storms that are far more typical of the tropics. They argue that this happened in October 1987, and in January and February 1990. This scenario is, however, unlikely as it presupposes that the temperature of the atmosphere remains the same while that of the oceans rises.

There is also a contrary view. Events such as these storms are generated by heat and, in a warmer greenhouse world, they are the sort of phenomena one might, therefore, expect. The paradox is that the greenhouse effect would make such severe storms less likely. In a greenhouse world, the greatest increase in warming would be in the polar regions rather than in the tropics (see p. 141) This would have the effect of reducing the differences between polar and tropical air, making the formation of deep depressions and severe storms less likely.

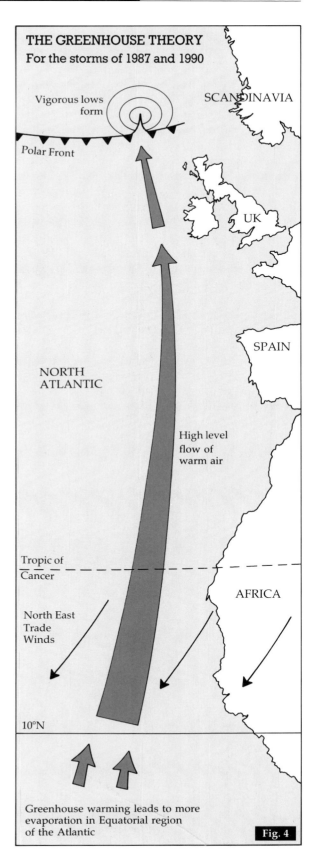

THE GREENHOUSE THEORY
For the storms of 1987 and 1990

Vigorous lows form

SCANDINAVIA

Polar Front

UK

NORTH ATLANTIC

SPAIN

High level flow of warm air

Tropic of Cancer

AFRICA

North East Trade Winds

10°N

Greenhouse warming leads to more evaporation in Equatorial region of the Atlantic

Fig. 4

IN FULL FLOOD

After a brief respite, the gales continued to batter the West of Britain with winds reaching over 180 km/h (100 mph). Five days after the first blast, a second wave of storms tore across Devon and Cornwall, Wales and the Severn Valley. Melting snow in Wales and heavy rainfall across much of western Britain brought widespread flooding, the worst for 40 years. At Worcestershire County Cricket ground the floodwaters reached the first floor balcony level and along the Dyfed coast, in West Wales, waves as high as nearly 10m (30 ft) inundated low-lying coastal pastures and drowned sheep.

The most dramatic flooding at the beginning of the year was at Towyn near Rhyl in North Wales. High waves were driven by the strong gales to breach the sea defences on the Clwyd coast and virtually the entire village of Towyn was flooded. Several thousand people were evacuated and those whose homes were most severely damaged were not able to return to the village for several months.

The whole of the flat coastal plain in the estuary of the River Clwyd was inundated and the main North Wales railway line was out of action. Pictures of the flooding at Towyn seemed to emphasize the warnings of a rise in sea level that might come about if serious global warming was to occur (see p. 128).

These floods affected widely separated parts of Britain, and there was much speculation that the flooding was the worst since 1947 and perhaps the most widespread this century. The Thames and the Severn were spectacular—where once the rivers flowed between winding banks there were miles of new 'lakes'. The Wye and the Usk burst their banks in the Welsh Borders and so did many of the shorter rivers of South Wales. The Exe and its tributaries flooded in Devon, while in Scotland the Ness and the Spey rose suddenly and burst their banks. Many towns were inundated, and among those most seriously hit were Inverness, Worcester, Hereford and Shrewsbury.

The severe flooding was the result of melting snow and of very heavy downpours which transformed the precipitation record for the winter of 1989–90, which started out with drought and ended, in many places, as being the second wettest winter of the 20th century.

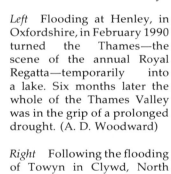

Left Flooding at Henley, in Oxfordshire, in February 1990 turned the Thames—the scene of the annual Royal Regatta—temporarily into a lake. Six months later the whole of the Thames Valley was in the grip of a prolonged drought. (A. D. Woodward)

Right Following the flooding of Towyn in Clywd, North Wales, during the storms of February 1990, plans were drawn up to strengthen the sea walls to protect this low-lying district from any future inundation. However, a rise in sea level that could result from global warming would submerge Towyn, and many other coastal towns in Britain, for good. (See map 139) (Mercury Press and Gamma)

FEBRUARY GALES

On Wednesday 7 February 1990, the third great destructive storm in just over two years swept across Britain. Although the gale reached a speed of 173 km/h (96 mph) at Brixham in Devon, the February storm did not reach the same ferocity as the great wind of 25–27 January. There were fewer fatalities, but trees were blown down, buildings weakened by the storm days previously were damaged and road and rail communications were disrupted once again.

Much flooding followed the storm. Whole villages were cut off, livestock drowned and considerable damage caused. The Severn valley was the worst affected, but the Thames also broke its banks in Berkshire and the Salisbury Avon held its highest flow since 1947, but the flood defences just managed to hold.

NORMAL?

The storms of January and February 1990 in Britain were not unprecedented according to the British Meteorological Office and the London Weather Centre. Blocking (see p. 25) held the pattern of the atmospheric conditions for several weeks and the pressure systems causing the gales that would normally have moved on hung around, preventing any substantial changes from being made.

The two weeks following the storm of 25 January saw the windiest spell for 30 years, but although the weather was extreme, it was not abnormal. There is great natural variability in the weather. Storms of this magnitude occur about half a dozen times each winter associated with deep lows moving East from off the North Atlantic. However, these deep depressions normally pass much further North, between Scotland and Iceland, and rarely reach their greatest intensity over the land. They usually bring severe gales to sea areas Bailey, Faeroes and South-East Iceland far to the North of Scotland.

So, if the conditions were not unusual, the location in which they were experienced was. Professor Brian Hoskins of the University of Reading commented that 'it's extremely difficult to know whether these are events at the edge of the normal range or the result of climate changes'.

Professor Hubert Lamb of the Climatic Research Unit at the University of East Anglia noted that the ferocious gales, yet one more example of climatic extremes, were consistent with the possible effects of global warming. At the same time he drew attention to a similar series of severe storms that were experienced in Britain during the 1890s. It has been suggested in some quarters that the gales of the late 19th century and the gales of late 20th century are linked as features of a cyclic weather pattern caused by solar activity (see p. 170).

Left The floods in Inverness in February 1990 proved not to be as devastating as those experienced exactly one year previously when the railway bridge over the River Ness collapsed, disrupting communications with the North of Scotland. (Met Office)

Right A computer-generated perspective view of a 'real' hurricane—Hurricane Allen, seen over the Gulf of Mexico on 8 August 1980. The image was produced from two satellite images, one taken in visible light, the other in infrared. The deep depression at the centre of the storm is dramatically illustrated. (Science Photo Library)

TORNADO

During the night of Tuesday 15 August 1989 a 162 km/h (90 mph) tornado devastated Butlin's holiday camp at Pwllheli in North Wales. The gust lasted less than 60 seconds but during that time it caused damage totalling nearly two million pounds. It advanced with a rumble resembling an express train and picked up trees and panels from the roofs of chalets in its path.

There were ready comparisons with the hurricane of October 1987, but a tornado is a very different phenomenon. Caused by violent local inequalities in air temperature and pressure, a tornado can in theory be caused anywhere over land when temperatures are high. It is a fiercely rotating column of air that is quite small in diameter—often as little as a couple of metres wide. Air is drawn up in the column which often does not reach down to the ground, but when it does—as was the case at Pwllheli—the damage can be severe.

Despite the dramatic nature of what happened that summer's night in North Wales, there is nothing particularly unusual about a tornado in Britain. On a small scale they occur every summer and 'mini-tornadoes' are the favourite theory used to explain the mysterious circles that appear in fields of wheat and other cereals in southern England.

3 A warm spell

The winter of 1988–89 was the mildest in England since records were first kept on a regular basis (in 1659 in some parts of the country or 1869 in Central England). Although this may have been welcomed by householders, anticipating lower heating bills, the news was seized upon by the press as further 'evidence' of global warming. *The Daily Telegraph*, for instance, sported the headline 'Global warming worries heightened by mildest winter for 330 years'.

Most scientists remained more cautious. The figures revealed that the winter of 1988 was some 2.5°C (36.5°F) warmer than the average for winters over the period 1961–80. However, a handful of recent milder winters cannot be interpreted as the start of the greenhouse effect. It is possible that we are living through a short natural variation in the climate, even though the cumulation of evidence from various sources leads many scientists to suggest otherwise. A growing number of observers believe that yet another mild winter is 'consistent' with various predictions of global warming.

A study of the winter of 1988–89—*The Mild Winter 1989–90*, published by the Institute for Terrestrial Ecology, which is based near Edinburgh—provides an indication of the sort of changes that might occur if global warming were to become a serious reality. The effects of one particularly mild winter alone had measurable consequences on British agriculture, fauna and flora.

Many crop and animal parasites are usually killed off by the cold winter weather. In 1988–89 significant numbers were able to survive the winter, causing increases in illnesses among livestock. There was a large increase in the aphid population and outbreaks of pests caused considerable damage to conifers, particularly Norway spruce. Higher temperatures allowed some animals to remain active for longer during the winter months, and although the activities of hedgehogs and bats may have been welcome, 1989 witnessed an explosion in the number of slugs.

The warmer weather promoted earlier growth of pasture, but although this early grass might have looked like a bonus, its lush appearance was deceptive. These pastures proved to have little value as livestock feed because they lacked important trace elements.

These are small changes noted in one milder winter. It is pertinent to ask how much greater the effects would be if climatic change were a reality. (See p. 106.)

It has been suggested that milder winters may have certain advantages to agriculture in Britain. Warmer conditions have permitted the increasingly successful cultivation of maize and of sunflowers in the southern counties of England. Harvests, too, have been earlier. Over the past few years the sight of combines, busy in the fields in July has not been uncommon South of a line drawn from Dorset to Norfolk. There is, however, another side to the coin. An early harvest is no guarantee that the grain will be as large as normal.

WARMING

In 1989, the Meteorological Office classified all four seasons as 'very warm'. This classification was unique in the 20th century. There was more sunshine in every month of the year than in any year since 1909. Even 1949—the warmest year this century—had less sunshine. Some people were quick to attribute the high temperatures of 1989 to global warming and the greenhouse effect; others pointed out that increased solar activity was an equally possible cause. The Met Office would not be drawn to any conclusions but did draw attention to the eccentricities of the climate. Yet despite 1989 being the second warmest year on record, the summer heat wave was not nearly as persistent as the hot spell in 1976 when there were 15 consecutive days on which temperatures passed 32.2°C (90°F). In the summer of 1989, the temperature only exceeded that level for a maximum of three days running.

Previous page The sunflower, *Helianthus annus*, a native of the Americas—grown for its oil—is only one of the 'new' crops that are gaining in popularity in southern England. Long hot summers have encouraged more farmers to adopt new crops from warmer regions. (Spectrum)

Over page A forest fire. Unusually hot summers dried forests to a tinder, causing fires in England, France and (here) in the USA during the droughts of 1988, 1989 and 1990. (Gamma)

1989—A FORETASTE OF A WARMER CLIMATE

It was Britain's second warmest year since the measurement of temperatures began in 1659. In much of England—and particularly in the Midlands and East Anglia—it was one of the sunniest years this century. On 22 July, at the Surrey village of Mickleham in the Mole Valley the temperature reached 34.4°C (93.9°F), the highest individual temperature recorded since 1976.

Apart from the long warm summer, 1989 will be remembered for the low rainfall totals received almost everywhere—the South and the East were particularly dry, although conditions were not as severe as during 'The Great Drought' of 1976—and for the remarkably mild winter months starting with a warm January.

January

Throughout the country temperatures were on average between 2° and 3°C (3.6° and 5.4°F) above normal. Scotland was outstandingly mild with temperatures about 4°C (7.2°F) warmer than usual. Except for the West Coast of Scotland, which was generally overcast and very wet, Britain enjoyed a much drier and sunnier January than normal.

Highest temperature:
  15.0°C (59°F)　　7 Jan　St Abb's Head, nr Berwick (Border)

Lowest temperature:
  −5.9°C (21°F)　18 Jan　Bastreet (Cornwall)
  −5.9°C (21°F)　19 Jan　Shawbury, nr Shrewsbury, (Shropshire)

Normal mean temperature (London):
  5.0°C (41°F)

February

Most of February was also extremely mild with temperatures in general 2° to 3°C (3.6 to 5.4°F above the seasonal average. However, the middle of the month witnessed another of the intense storms that some scientists think may be a sign of climatic change. Because the fierce gale struck the North of Scotland rather than the Home Counties, this storm did not receive the same blanket media coverage as the 'hurricane' of October 1987 or the violent gales of February 1990, yet at Fraserburgh (Grampian) on Thursday, 13 February a wind speed of 123

knots was recorded. This was the highest gust of wind ever measured at a low-level meteorological station in Britain. Low rainfall continued in the East of England and Scotland, but the West and the North were wet; Northwest Scotland was very wet.

Highest temperature:
  15.8°C (60°F)　　6 Feb　London Weather Centre

Lowest temperature:
  −7.1°C (19°F)　27 Feb　Aviemore (Highland)

Normal mean temperature (London):
  6.0°C (43°F)

March

Temperatures continued to be above average—by almost 3°C (5.4°F) in the southern and eastern counties of England. Rainfall totals were closer to the seasonal normal than in the preceding months, although the Northeast was drier and the West of Scotland wetter than usual.

Highest temperature:
  19.9°C (68°F)　28 Mar　East Hoathly, nr Uckfield (East Sussex)

Lowest temperature:
  −7.4°C (19°F)　17 Mar　Tummel Bridge, nr Pitlochry (Tayside)

Normal mean temperature (London):
  7.0°C (45°F)

April

If there was more than a hint of spring about March, April proved to be both colder and wetter than normal. During the first week of April a short-lived carpet of snow covered parts of the Midlands and the Thames Valley. Snow fell again in many places between 22 and 25 April, and Nottingham woke up to 100 mm (nearly 4 in) of snow on 25 April. In the middle of the month gales hit western districts and caused structural damage in Wales.

Highest temperature:
  17.7°C (64°F)　30 Apr　Cromer (Norfolk)

Lowest temperature:
  −7.8°C (18°F)　25 Apr　St Harmon, nr Rhayader (Powys)

Normal mean temperature (London):
  10.0°C (50°F)

May

The Sun made a welcome return in May. Most places basked in over 300 hours of sunshine and the summer seemed to have arrived. In the South of England temperatures were up to 4°C (7.2°F) above the seasonal average, but the warm weather also brought the first hints of drought. The month was unusually dry, particularly in eastern and southern districts, although occasional violent thunderstorms, such as the one at Woking (Surrey) on 24 May, brought heavy rainfall. Overall, though, it was one of the eight driest months of May since rainfall was first measured on a regular basis in Britain in 1727.

Highest temperature:
29.4°C (85°F)   23 May   Heathrow Airport
29.4°C (85°F)   24 May   London Weather Centre

Lowest temperature:
−2.6°C (27°F)   10 May   Tummel Bridge, nr Pitlochry (Tayside)

Normal mean temperature (London):
13.0°C (55°F)

June

Although cooler, showery spells marked the start and the end of the month, June was mainly warm, fine and remarkably dry. Rainfall was variable, with Merseyside being wetter than average. In Somerset, Avon, Dorset, Devon and Cornwall rainfall totals were much below the normal. The water authorities began to worry, but most people were content to enjoy another month of sunshine.

Highest temperature:
30.5°C (87°F)   20 Jun   Leeds Weather Centre (West Yorkshire)
30.5°C (87°F)   20 Jun   RAF Finningley, nr Doncaster (South Yorkshire)

Lowest temperature:
−1.1°C (30°F)   4 Jun   Cellarhead, nr Stoke-on-Trent (Staffordshire)

Normal mean temperature (London):
16.0°C (61°F)

July

Sunshine and temperature records were broken throughout July. The 22nd was the hottest day recorded since 1976. All over Britain temperatures were on average 2–3°C (3.6–5.4°F) above normal and the combination of heat and sunshine brought the harvest forward by more than a month in southern districts. The dry period continued. Thunderstorms brought rainfall to parts of East Anglia, but most of the rest of the country lacked adequate rain and restrictions on the use of water came into force in many places.

Highest temperature:
34.4°C (94°F)   22 Jul   Mickleham, nr Dorking (Surrey)

Lowest temperature:
2.2°C (36°F)   11 Jul   Eskdalemuir, nr Langholm (Dumfries & Galloway)

Normal mean temperature (London):
18.0°C (64°F)

August

In August the weather broke in Scotland and the North which became cool, cloudy and wetter. Parts of the North West were rainier than normal, and North Wales was on the receiving end of a freak tornado which wrecked Butlin's holiday centre at Pwllheli in Gwynedd (see p. 43). The rest of Britain, however, continued to bask through the long, hot, sunny summer.

Highest temperature:
30.4°C (87°F)   20 Aug   Coltishall, nr Norwich (Norfolk)

Lowest temperature:
0.9°C (34°F)   1 Aug   St Harmon, nr Rhayader (Powys)

Normal mean temperature (London):
18.0°C (64°F)

September

Although September was hot and sticky, it tended to be overcast rather than sunny. Temperatures remained above normal, especially in southern counties. In many places this was the driest month of an exceptionally dry year.

Highest temperature:
27.2°C (81°F)   7 Sep   RAF Benson, nr Wallingford (Oxfordshire)
27.2°C (81°F)   21 Sep   St Helier (Jersey)

Lowest temperature:
−3.1°C (38°F)   11 Sep   Aviemore (Highland)

Normal mean temperature (London):
16.0°C (61°F)

October

Over part of the country the drought continued. East Anglia received unusually low rainfall but Wales, the West Country and western districts of Scotland experienced normal and in some cases above average rainfall. October was a warm month, but the last two weeks were changeable and increasingly windy. Strong gusts in the South and West did not reach the ferocity of The Great Storm of October 1987.

Highest temperature:
22.3°C (72°F)   5 Oct   Herne Bay (Kent)

Lowest temperature:
−2.6°C (27°F)   2 Oct   Tummel Bridge, nr
                         Pitlochry (Tayside)

Normal mean temperature (London):
13.0°C (55°F)

November

What was possibly the sunniest November this century reinforced, in some minds, the idea that the effects of global warming were already being felt. Some South Coast resorts recorded over 100 hours of sunshine for the month. The pattern of low rainfall continued, with many places receiving under half their normal November fall. The mild weather lasted until the middle of the month when temperatures returned to the seasonal normal and, in some areas, were lower than usual.

Highest temperature:
18.9°C (66°F)   13 Nov Nantmor, nr
                        Porthmadog
                        (Gwynedd)

Lowest temperature:
−9.4°C (15°F)   26 Nov St Harmon, nr
                        Rhayader (Powys)

Normal mean temperature (London):
9.0°C (48°F)

December

After a week of dry weather, the long-awaited rain came in December. In most southern districts this proved to be the wettest month of 1989—London, for example, had its two wettest weeks in December. In northern England and in Scotland much of the precipitation came as snow, with quite heavy falls in some places. Exposed western coasts were battered by severe gales—a foretaste of the intense storms that were to follow in January and February 1990.

Highest temperature:
15.0°C (59°F)   20 Dec Skegness
                        (Lincolnshire)

Lowest temperature:
−13.1°C (8°F)   1 Dec Tummel Bridge, nr
                        Pitlochry (Tayside)

Normal mean temperature (London):
6.0°C (43°F)

What does the experience of one year prove? In climatic terms, not very much. The long, hot summer, the low rainfall and the increased incidence of severe gales during an otherwise mild winter can be noted from the survey of the weather of 1989. A study of the figures for other recent years would, in part, reveal a similar trend. But three, four or five years is not long enough even to acknowledge a trend. These changes could be temporary, or they could be longer lasting. It is also possible that the unusual similarities in weather could be coincidence. We should not expect the weather to remain the same, but we do not yet have sufficient evidence to prove that the climate is changing.

IS IT REALLY GETTING WARMER?

There is an assumption that temperatures in Britain are increasing. The record temperature of 37.1°C (98.8°F) at Cheltenham on Friday, 3 August 1990, received a very great deal of publicity and there was much speculation that the temperature might reach the 'magic' figure of 100°F in that heatwave. But for many years the Meteorological Office accepted a temperature of over 100°F as the British record. On 22 July 1868 a temperature of 38.1°C (100.5°F) was recorded at Tonbridge in Kent but that 'record' has since been discredited because it was taken on a 'non-standard' screen.

The only temperatures now officially accepted are those recorded in a Stevenson screen which has a thermometer inside a box. Shade temperatures are recorded rather than the artificially inflated ones that would be obtained if the instrument were to be left in the direct Sun and subject to varying wind speeds. The box is raised on legs above the ground and has four louvred sides for ventilation. It is painted white as boxes of other colours will not reflect the heat so efficiently and the readings obtained from these would be inflated.

The Tonbridge figure may not be regarded as an official record now but there can be no doubt that July 1868 was exceptionally hot for Britain, as newspaper reports of the time confirm. Other particularly high temperatures were recorded in 1911 and 1932 at a number of British sites. The reading that was until 1990 the official record—36.7°C (98°F) at Raunds in Northamptonshire—was only one of several exceptional figures obtained on 9 August 1911. At Islington in North London, 37.2°C (99°F) was recorded and although it was read in a Stevenson screen, it was not accepted by the Meteorological Office—even though it was an official Royal Meteorological Society site. (The reason for discounting this figure has never been given.) On the same day, 37.8°C (100°F) was recorded at Greenwich, again in a Stevenson screen and again not officially recorded.

It should be remembered that these were only the *recorded* figures. It is probable that these temperatures were exceeded in places that were not fortunate enough to have weather stations.

We should not be persuaded that 'weather records' are progressive—or because records are being broken almost continually in one way or another, that it is getting hotter, drier, wetter, windier, colder or whatever. Records are being broken because more recordings are being made in more places worldwide. We are building on a relatively short time-span during which weather stations have been established, and an even shorter time-span during which accurate readings, using reliable equipment, have been possible. It should come as no surprise that readings above and below those recorded since the turn of the century are continuously being taken.

We have figures, therefore, that indicate that 1911 and 1932 were as hot as 1989 and 1990 in Britain, and early records—all based on data obtained from equipment that would not be accepted today—suggest at least four exceptional heat waves in the 19th and 18th centuries in Britain.

The great expert on early British weather figures was Professor G. Manley who believed that in the month of July in 1757, 1808, 1825 and 1868 there were heat waves in southern England during which temperatures exceeded 35°C (95°F). One day in July 1808—the 13th—entered English folklore as Hot Wednesday. During the second half of the 19th century

Left Temperature figures used in climatology are almost always temperatures in the shade, that is the temperature of the air taken in conditions that rule out the effects of the direct sun. Temperatures are normally recorded in the controlled conditions inside a Stevenson screen. The highest shade temperature ever recorded was 58°C (136.4°F) at al'Aziziyah in Libya on 13 September 1922. (Met Office)

Right In the exceptional drought of 1990, reservoirs across southern and western Europe ran low. This reservoir in the Lauragais area of southwest France supplies the city of Toulouse with drinking water. As drought orders were enforced, some villages were obliged to rely on stand pipes. (Gamma)

many old countrymen entertained their listeners with stories of how rivers ran dry, crops shrivelled and cottages subsided as the ground cracked in the summer of 1808, and how on Hot Wednesday itself animals collapsed and died of heat in the fields. Stories from newspapers of the time read remarkably like those in 1989 and 1990—only the style in which they appear is different.

The past reveals other similar years—1707 and 1513 also entered folklore. From descriptions of the conditions then, it is possible to suggest that the weather in the summers of 1513 and 1707—and in several other years—resembled those experienced recently in Britain.

WARM WINTER

As well as being notably windy, January 1990 was one of the two warmest first months of the year in Britain since records were first kept in 1659. The other was January 1773 when average temperatures were exactly the same as in January 1990. And 1773 was, of course, long before anyone began talking about the greenhouse effect. From past climatic records it is possible to identify a number of similar months which are evenly spaced out over the last three centuries, suggesting that such 'abnormalities' are a regular occurrence, an extreme at the edge of the normal range of climate. In 1733, 1793, 1834, 1916 and 1921 the month of January was mild and warm.

In southern Britain the temperature reached over 15°C (59°F) on 11 January 1990, but towards the end of the month the area of high pressure over Europe that was responsible for the mild conditions was energetically pushed aside by a very vigorous low in the North Atlantic. This deep depression brought the fierce storm that howled across much of Britain between 25 and 27 January (see pp. 38–39).

DROUGHT

If significantly greater precipitation and flooding has been predicted for some regions of the world as a result of global warming, the reverse is forecast for many other areas. From newspaper and television news reports of the drought in Ethiopia and the Sahel we have a graphic picture of the results of drought, and may wonder if such conditions could be experienced in other regions by the middle of the 21st century as a result of climatic change.

In Britain the droughts of the summers of 1989 and 1990 were seized upon by some as an early sign of the greenhouse effect, but the long dry spell was especially abnormal and could equally well be explained in other ways.

The prolonged British drought of the summer of 1990 had its roots in a long spell of unnaturally dry weather during the previous autumn and winter. In some parts of the country, the low rainfall of much of the winter did not allow reserves in the ground and in reservoirs to recover after the dry conditions of 1989, which in Britain was the second driest year in more than a century. In Southeast England the long dry spell stretched back as far as April 1988.

Are there droughts, forerunners of climatic change, in other parts of the world? Looking back over the last three years it would be possible to find newspaper stories concerning drought not only in the Sahel but also in Australia, in the United States, China, Central America, Namibia and South Africa, and in parts of India, Brazil and Turkey. A sign of global warming? Not really, because similar stories could be found concerning a lack of rain in the same areas during almost any decade

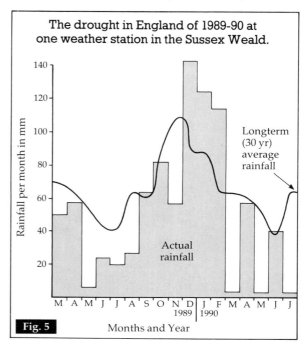

The drought in England of 1989-90 at one weather station in the Sussex Weald.

Rainfall per month in mm

Longterm (30 yr) average rainfall

Actual rainfall

M A M J J A S O N D | J F M A M J J
1989 | 1990

Fig. 5

Months and Year

this century. In reply to the increasingly common suggestion that the rainfall totals received in arid zones are decreasing, it can be said that precipitation in such arid regions as South Africa and Western Australia is notoriously unreliable. The rainfall figures for most cities in Nevada, USA, for example, reveal considerable variation from one year to the next. It should be remembered that when we speak of 'average rainfall' we are referring to an average which by international convention is calculated over a span of 30 years. The 'average' is seldom, if ever, received, but rather a total above or below that figure. Even years of drought may not mean a change in climate—the likelihood is a period of rainfall well below the average. In the Sahel, for example, the continuing drought was dramatically interrupted in some places in the South of the region by well above average rainfall in August and September 1988, allowing the first good harvest for very many years to be gathered (see p. 14).

DRINKING WATER

In parts of Southeast England drinking water is taken from the chalk hills of the North and South Downs and the Berkshire and Hampshire Downs. It is stored in natural cracks, deep underground in the chalk which acts as a giant sponge. The water is then pumped out for use or obtained from springs where it comes bubbling from the ground at the junction of the chalk and the clay beneath it. But with little rain for months, the giant sponge of the chalk downlands began to run dry in the summer of 1990.

Throughout southern England, the level of streams fell. Some streams of the chalk lands are more or less seasonal anyway, but even well-established rivers suffered as the drought continued. Few counties were as badly affected as Berkshire. The seasonal streams—the winterbournes—were bone dry. Some bournes were reduced to cracked and dusty dips where no water had flowed for months. The downlands of Berkshire are crossed by small green valleys through which trickle willow-hung streams which are occasionally blocked to form watercressbeds. In the long hot summer of 1990 those valleys were parched yellow and brown, the streams had run dry and the watercress beds were largely disused.

Water companies were forced to extract more and yet more water from the rivers and the porous chalk downlands, further reducing the flow of the rivers. In Berkshire the River Pang, a famous and beautiful trout stream, dried completely for part of its course, and the headwaters of up to a dozen other substantial rivers disappeared completely in the Thames Valley.

CLIMATIC CHANGE?

The long hot summers of 1989 and 1990 brought sultry holiday weather to southern Britain for week after scorching week. Sun worshippers perfected their tans, bought cool drinks and wondered just why they had bothered to book a holiday in Greece when the weather in Britain was so good.

Not only were the summers hot, they were also abnormally dry. In the chalk downlands of southern England, some streams eventually dried up. Gradually the level of reservoirs dropped to expose stretches of baking cracked mud to a Sun that seemed forever to be shining. Gardens were parched. The grass became dry, brown and brittle. The use of hose pipes and of car washes was banned and there were threats of stand pipes in the streets.

Harvest was early, but the lack of moisture meant that the grain was not large. Cuts of silage were poor and had it not been for a welcome fall of rain in September 1989 many farmers in the South of England would not have had any grass left to cut. The farmers would have given much to hear the sound of heavy rain again. Day after day of clear, dry weather may have given joy to holiday-makers but countrymen would have preferred to see water in the ditches and ground that was not as hard as rock.

Everyone agreed that 1989 and 1990 were not 'typical English summers'. But hadn't 1987 been unusually warm too? In Britain the summers of 1981, 1983, 1984, 1987, 1989 and 1990 all produced temperatures well above average. Was the climate changing? The newspapers were full of stories about 'global warming', 'climatic change' and 'environmental disasters', but what was the truth?

4 Climate past

Some commentators have ascribed the weather during the British summers of 1989 and 1990 to global warming and the greenhouse effect, yet the evidence from the past suggests that similar conditions have been experienced before and even seem to reoccur in a pattern. If greenhouse warming has been at work we would expect the weather to be getting progressively hotter and drier in the summer, but the weather record of the past two decades does not bear this out. The summer of 1976 was hotter over a longer period than the summers of 1989 and 1990; it was also substantially drier. Indeed, 1976 is still the driest year on record in Britain rather than, as might be expected, one of the more recent years during which global warming is suspected to have taken effect. (It could be argued that global warming was happening as early as 1976 for carbon dioxide levels were increasing then.)

The historical record of regular droughts and heat waves in Britain may not come up to the exacting requirements of modern scientific observers. Nevertheless, accounts of extremes of weather are a recurring theme in English history. The people of Sussex, for example, were said to have been starving as a result of three years of drought and hot weather when they were converted to Christianity by St Wilfrid at Bosham and Chichester in 681. We have, it seems, experienced before the sort of weather that many people are now so worried about, and 681, 1513, 1707 and 1757—all years of great heat and drought—were all long before the increased emission of greenhouse gases. It could, therefore, be suggested that regular hot periods of sustained drought are a normal part of the weather pattern in Britain.

The unusually hot summers of recent years in Britain—in 1975–6, 1981–4 and 1989–90—have come together in a way that some people have interpreted as an indication of a change in the climate. However, a cluster of warm summers is not unknown from the past climatic record.

CLIMATE PAST IN BRITAIN

The ecclesiastical historians writing at a later date condemned the godless and violent early Saxon days as the 'Dark Ages'. From the point of view of the weather, seldom can an age have been so inaccurately named, as the climate of Anglo-Saxon England was warmer and generally more pleasant than the conditions we experience now.

From chronicles, farm records, letters, war reports, tree rings, archaeology and many other sources of 'trapped' climatic data, we can build up the story of the climate during past periods of British history. The picture is one of constant and often sudden change, of a broad canvas of climatic trends in the long term, but also with the finer detail of short-term variations lasting sometimes a couple of centuries and on other occasions just a decade or two.

The whole is a pattern or cycle of change, whose discovery is comparatively modern. In the safe world of Victorian England, stability was a key, and any suggestion of climatic variation or change would have been generally discounted. The measurement of climatic elements was only just achieving modern standards of accuracy. Only after the Stevenson screen (see p. 51) was invented last century could any truly accurate recordings of temperature be made, and when these were available over any length of time, the variability of climate became evident.

It was, of course, known that the climate could not always have been the same. The coalfields were laid down in Lancashire and Yorkshire when those areas were in the deltaic swamps of a great tropical lowland. The flat open landscape of East Anglia was created by ice sheets, both by erosion by the ice and by the deposition of glacial debris eroded from elsewhere further North. These were changes on a grand scale, changes that had taken place long ago. Yet, using recorded data of rainfall and temperatures, the Victorians learned that there were lesser variations from the 'normal' weather on a smaller time scale.

The 'normal' is taken to be an average over a period of 30 years. It is this benchmark that is referred to when we read or hear that 'the temperature will be above average for the time of year' or 'rainfall last month was below the seasonal average'. The current average refers to

Right Ten frost fairs are known to have taken place on the frozen River Thames. This contemporary cartoon by George Cruikshank illustrates the final frost fair held in 1814. (Mary Evans Picture Library)

the period 1951–80, and this may lead us to treat one or two theories about widescale climatic change with a little caution. The 30-year-period that covered the 1950s, 1960s and 1970s was, on average, cooler and wetter than the 1980s. If we are currently accepting as 'normal' the conditions of a period which is dissimilar to our own, we should not be too surprised, or draw too many conclusions from the frequency with which temperatures are 'above average' or rainfall fails to reach the 'normal' amount for the time of year.

CLIMATIC EPOCHS

In the long term we can recognize over half a dozen periods of global climatic change, known as epochs, since the final retreat of the ice at the end of the last glaciation.

About 10 000 BC civilization began. Man became a settled farmer, built dwelling places, and began to live in societies more complex than the extended family. He soon found means of recording his history, his crop yields and religious beliefs. Since then, changes in climate have been recorded and can be deciphered in letters, archaeological finds, monuments, folklore and so on.

The first epoch, after the Ice Age, was warm with summer temperatures higher than those experienced now. Palaeoclimatologists have calculated from pollen records and tree rings that winters in that epoch were unusually mild throughout Northwest Europe and much of North America, with temperatures in the region of those suggested by some greenhouse theorists for the 21st century. This Pre-Boreal epoch lasted until about 4500 BC. Sea levels had risen with the melting of the ice and precipitation was high.

The Pre-Boreal epoch was followed by an even warmer period, the Atlantic epoch or Post-glacial Climatic Optimum. Then, around 2500 BC, the third modern climatic epoch—the Sub-Boreal—arrived, bringing a climate similar to the Pre-Boreal epoch, although there is some evidence of great fluctuations in temperature with occasional very low winter temperatures.

About 1000 BC, a colder epoch—the Sub-Atlantic—brought a fall in average summer temperatures of about 2°C or 3°C (3.5°F or 5.5°F), although the winters remained relatively mild and wet. This epoch, which coincided with the beginning of the Iron Age in Britain, brought strong winds. The colder spell lasted until about 450 BC when temperatures

increased again. The warming intensified and from around 1150 to 1300 a much warmer climate was experienced throughout the temperate latitudes.

This warmer epoch is now usually referred to as the Little Optimum. In North America, the Little Optimum brought warmer weather to some areas that had previously been marginal to Indian tribes that relied upon farming for their livelihood. These areas, for example in the Northwest states, were settled by Indians for the first time, although others were forced out of some southern locations owing to drought.

In Britain, agricultural changes were significant in the Little Optimum. Vineyards flourished as far North as Yorkshire and the level of cultivation in the fells of Northumberland rose to about 320 m (1000 ft). The spread of cultivation was so great that grazing land was squeezed out by arable farming in some areas. This, too, was the era of maximum Viking settlement in Iceland and Greenland (see p. 66 and 68). Scotland had a much milder climate than today during the Little Optimum, and it can be no coincidence that this early medieval period was one of considerable prosperity for Scotland which, briefly, became self sufficient in food.

THE LITTLE ICE AGE

The Little Optimum came to a rather sudden end around 1300, although in some places evidence of climatic change dates from as early as 1210. By 1320 all Northwest Europe was in the grip of what has come to be known as the Little Ice Age. Both winter and summer average temperatures fell by over 1°C (1.8°F) and the annual average temperature in Britain decreased by 0.5°C (0.9°F). These figures disguise the extraordinarily low temperatures experienced during some winters which earned this epoch its very graphic name. The abruptness of the change can be seen by the rapid abandonment of many vineyards, particularly in Germany where viticulture had to move up to 200 m (600 ft) down slopes in order for the grapes to ripen. In England, vineyards virtually disappeared.

The violent storms of the Little Ice Age became legendary. Scottish farming declined as temperatures decreased and Scotland passed in 100 years from prosperous self sufficiency to famine and poverty. In 1270 the Thames in London froze, in part because the new London Bridge, with its narrow arches and broad piers, greatly impeded the flow of the river.

Left Drusilla's Vineyard at Alfriston in Sussex is one of the growing number of English vineyards. Encouraged by warmer summers, viticulture is advancing further North. In 1990 the most northerly vineyard was thought to be at Sabile in Latvia, which is just North of 57 N. (Spectrum Colour Library)

Right Grindelwald Glacier, Switzerland. Since 1850 there has been an overall shrinkage of the glacier as mean temperatures have increased. The rate of recession has been sporadic with a retreat of over 70 m (210 ft) in 1961–62 alone. Over the past 30 years, the retreat of the Grindelwald Glacier has revealed tree trunks that were overwhelmed by the advance of the ice in the later Middle Ages. (Images)

The Little Ice Age destroyed the Norse settlement of Greenland (see p. 68) and nearly wiped out the population of Iceland (see p. 67). But all was not wind, snow and ice during this prolonged epoch. In the 15th century in Britain there was a brief warming when 1433, 1434 and 1436 all had remarkably warm summers, the descriptions of which in the chronicles sound much like the long hot summers of 1989 and 1990. In many ways the 1430s must have been a very confusing decade to live through. Not only were they easily the warmest in living memory, they were also the wettest. The abnormally rainy years of 1432, 1437, 1438 and 1439 witnessed flooding of unprecedented severity, and the biting cold of the winters of those years added to the misery of crop failure in England. It is probable that the abnormalities of the weather from 1432 to 1439 were greatly in excess of anything experienced in Britain in the later 1980s when concern about climatic change was aroused.

The abrupt advent of colder wetter weather brought major changes to agriculture and to population distribution in Europe. Sites that were profitably habitable during the Little Optimum became unattractive in the Little Ice Age and between 1320 and 1360 many villages throughout Britain and Northwest Europe were abandoned. Settlement and farming retreated in 14th-century England and, traditionally, the Black Death of 1348–50 has been blamed. However, from surveys of tax records it has been shown that the major decline in the rural population in England in the 14th century was in the first 40 years of the century, that is *before* the arrival of the plague. This decrease correlates with long, harsh winters and late springtimes, followed by poor, wet summers. Crop yields fell drastically. What famine, induced by the climate, began, the Black Death finished off. A weakened population submitted to the ravages of disease and many village sites were finally abandoned. Many more English villages were also deserted in the middle of the 15th century when the plague was not a serious factor but the weather was especially poor.

There was a brief respite from 1500 to 1540, but the worst of the Little Ice Age was yet to come. The period 1550 to 1700 was the coldest in Britain, but the Little Ice Age did not finally end until the middle of the 19th century. Yet even during the most severe times of the Little Ice Age there were occasional warmer summers with good harvests. The mid-1660s were such a warmer interlude when drought, rather than heavy rainfall, was the rule. The Plague of

1665 was encouraged by the hot and dry weather experienced that year, while the tinder dry conditions of the following dry year meant that London burned more readily in the Great Fire of 1666.

That hot year of 1665 must have been another perplexing year. It began with a particularly cold snowy winter during which the Thames froze, an event that happened six times between 1407 and 1665. The thickness of the ice on the river was enough to support the weight of coaches and carts that were regularly driven onto the ice that year. (In the reign of Queen Elizabeth I, the 'Virgin Queen' was able to lead her court across the frozen Thames on a stately walk.) Frost fairs on the ice-choked rivers became a regular feature in the 17th century. Booths and sideshows, beer tents and dancing bears took to the ice, and in 1662 ice-skating was introduced to England for the first time and became the fashionable craze of that winter's frost fair.

In 1683–84 it would have been possible to skate on the sea when many coastal inlets along the English Channel and the North Sea froze. Diarists of the time recorded a heavy toll of sea birds, woodland birds and livestock as creatures froze to death. Cattle even perished of cold in their byres. During that bitter winter, the ground is said to have frozen to a depth of over a metre (nearly four feet).

The coldest year of the Little Ice Age in England was perhaps 1607–08, which has gone down in folklore as 'The Great Winter'. So many livestock died that hunger was widespread. Trees died of frost and others were split right down their trunks by great cracks expanded by the ice. In Scotland conditions were worse. When the Firth of Forth froze, the curious were able to walk out over the frozen sea to ships trapped in ice floes off the coast by Fife and Edinburgh. And right across Europe 1709 and 1830 were also especially severe years.

Right across Europe, the climatic belts seem to have shifted South during the Little Ice Age. Some areas of southern Europe exchanged the problems of drought for difficulties caused by frost, torrential rainfall and overcast skies. After 1280 the Grindelwald glacier in the canton of Bern (Switzerland) moved forwards, uprooting trees, and achieved a maximum advance in 1600, when it stabilized. (Since the

19th century the Grindelwald glacier has retreated to such an extent that the hotel built to offer visitors a spectacular view of the glacier is now far from the ice whose snout is much further up the valley.) The rate of advance of the glaciers in Scandinavia was greater during the Little Ice Age, with some farmland disappearing under the ice in central Norway.

The climatic record of North America parallels that of Europe during this epoch. The Pilgrim Fathers faced some of the coldest winters of the Little Ice Age during their first few years in the New World. In the southern hemisphere, however, conditions seem to have improved. The 'cone' of South America appears to have been warmer during the Spanish colonial period than it is now. Further South, the small ships of 17th- and 18th-century explorers penetrated deep into the seas around Antarctica and from their accounts the waters were freer of pack ice than they are now.

MODERN CLIMATIC EPOCH

In the 18th century, brighter spells of weather were more frequent, and some architectural historians have suggested that the large windows typical of Georgian houses were intended to benefit from the sunnier summers experienced after about 1730.

The Thames froze over for the last time in 1814 and, from the first quarter of the 19th century, temperatures in Britain increased. The modern climatic epoch has been one of considerable warming. Temperatures rose appreciably during the 1930s and 1940s, and in the period 1925–55 average temperatures were over 1°C higher in Britain than in the period 1895–1925. The drier, warmer weather of the 1930s encouraged the building of houses with flat roofs which became very popular in the decade preceding World War Two. Winters became milder and 'white Christmases' seemed to be confined to 'Dickensian' Christmas cards which still faithfully record the colder snowy winters of the 18th century.

The British climate this century—the climate that we regard as 'normal'—can be seen, therefore, to be quite unlike the climatic conditions experienced over the previous 300 to 400 years. From this brief survey of the climate in Britain over the past 10 000 years it is obvious that conditions are constantly changing. We are now in the warmest epoch since the early medieval period and the natural trend still seems to be towards warming.

The emission of greenhouse gases will also lead to warming, and we are faced with the puzzle of how much, if any, of the increases in temperature of the 20th century are due to the greenhouse effect. It may be that *all* of the warming so far is due to natural causes. In the present and the future we may be faced with a situation in which natural warming is added to by artificial, or greenhouse, warming, possibly with disastrous consequences.

PALAEOCLIMATOLOGY

The study of past climates over even longer periods may hold important clues to understanding the processes involved and to forecasting future trends. A study for the Royal Society *Frontiers of Earth Sciences to the Year 2000*, which was published in 1989, stressed the relevance of research into the flora and fauna of past periods through studying fossils. Palaeoclimatology suggests that climatic change in the past—for example a switch from a temperate climate to a polar climate—was rapid with certain species disappearing suddenly. This suggests climatic change over a period of decades or even few years rather than a gradual alteration of conditions over centuries.

There is evidence of great instability in the Earth's climates in the past one and a half million years. Palaeoclimatologists have been able to identify at least 17 periods during which the climate of the temperate latitudes was much colder than it is today.

Differences in the fossil record could be interpreted as a natural cycle of irregular fluctuations in climate. We have firm evidence of climatic change in the past—natural change in which man played no part. There is no reason to suppose that that natural cycle is no longer active, but what we do not know is whether human activity could intensify a natural warming or whether a cyclic cooling might dampen down a man-made greenhouse effect.

Fashions in building may reflect changes in climate. The construction of the Georgian and Regency terraces at Bath show the revival of interest in the Classical architecture of Greece and Rome in the 18th century. It has been suggested that the large windows of Georgian architecture were, in part, a response to the particularly sunny summers that characterized the period. (Spectrum Colour Library)

PAINTINGS

Paintings form a remarkable record of climatic change. European cave paintings from the Stone Age depict animals that live in warmer climes, while there are cave paintings of mammoths from another climatic era to be found in France. Ancient Egyptian murals show lions, elephants and other creatures that now live much further South in Africa, but which lived in a wetter Egypt in the past.

Even allowing for some artistic licence, the Flemish and Dutch painters of the 16th century depicted harsher, snowier winters than those of the 20th century. There can be few more graphic depictions of the conditions of the Little Ice Age than *Hunters in the Snow* by Pieter Bruegel (*c.*1525–69). Bruegel's scientific observation in many of his paintings is remarkably accurate, for example in his ships and architecture. There is, therefore, little doubt that his depiction of severe winter conditions in a number of paintings was true to life.

5 Coping with climate change

We have seen in the previous chapter that climatic change is nothing new and that the conditions that we are currently experiencing are short-term, since the end of the Little Ice Age in Europe, although we do not know how long they will last into the future.

Predictions of climatic change—as a result of global warming—abound. We are faced by a bewildering array of forecasts and warnings that our lifestyle may change (see p. 128 to 140). Our ancestors, too, were faced with sudden changes in the type of weather they experienced. How did they cope? In this section we shall look at several examples, taken from Europe, North America and the Middle East.

ICELAND

The island republic of Iceland is remote in Europe, and if its inhabitants feel that they are living on the edge, their intuition is justified. In past centuries natural disasters have made their country a dangerous place in which to live and in between the Middle Ages and the start of the 19th century, climatic change devastated Iceland's economy.

Because of the effects of the Little Ice Age, the Icelandic climate deteriorated during the 13th century and temperatures remained cool well into the 19th century. So great was the change that previously fertile farmland was lost to the advance of glaciers.

Breidarmerkur on the Southeast coast, for example, was a land of fields and woods, a rich farming area that had been settled since at least the 10th century. The valley of Breidarmerkur—literally 'broad boundaries'—enjoyed warmth because of the reflection off nearby glaciers. For centuries the area prospered, even through the great volcanic eruption of Oraefajökull in 1362 which covered part of the valley with ash. But in the cooler 18th century the Breidarmerkurjokull glacier advanced, covering the farms with ice and reaching to within 200 m (660 ft) of the sea. In the 20th century the climate has

improved and, with the warming, the glacier has slowly retreated, but the fertile green valley is no more. The melting ice has left behind Breidarmerkursander—the sands of Breidarmerkur—a wasteland of glacial debris.

Cycles of warming and cooling have had a considerable effect upon Iceland's history. The early history of Norse settlement on this volcanic island is one of adventure. Its stirring legends were captured in the Iceland sagas. But after 1262—when Iceland lost its independence to Norway—a long period of decline set in. Not only did the spark seem to go out of the country's literature but also a series of disasters dogged settlement and prosperity. Trade and industry passed under foreign control, while violent volcanic eruptions, famine and endemic cattle disease reduced the Icelanders

The Icelandic coast in January. During the 18th century such conditions lasted for most of the year, and the sea between Iceland and Greenland froze, allowing a few polar bears to cross into northern Iceland. (The Hulton Deutsch Collection)

to dire poverty. This was accentuated by a notable cooling during the 13th century, as western Europe began a long, cold period—the Little Ice Age (see p. 60).

The agriculture that had sustained the first Icelanders struggled to come to terms with colder winters, increased rainfall, shorter summers and the advance of glaciers. Climatic change meant that the growing season became dangerously short and in some places land fell out of cultivation. Sheep and cattle were unable to find forage and large numbers of livestock died.

The decline in the country's agriculture resulted in hunger and a sizeable drop in its population, while the disappearance of the majority of Icelandic sheep destroyed the nation's trading basis. In the 20th century the mainstay of the country's economy is, of course, fish. It may come as something of a surprise to learn that in the past Icelanders depended upon wool. Between the 11th and the 13th centuries the export of locally-spun woollen cloth was Iceland's economic foundation.

With only a short respite, the Icelandic climate remained much colder than at present, until the 19th century. The Little Ice Age reached its peak around 1750 when the lowest temperatures were recorded in Iceland. In the middle of the 18th century snowfall was common in Iceland up to June, quite frequent in July and was even experienced in August. As a result agriculture was rendered almost impossible, with only a very short grazing season and little opportunity for hay making.

The country reached depths of economic misery. With the Greenland Sea perpetually frozen and the rest of the coast icebound for all but a few weeks of the year, fishing was confined to the increasingly short summer. The coldest years were 1753 and 1754, when farm animals froze to death. The population declined to around 50 000 and when a serious smallpox epidemic, severe flooding and a major outpouring of sulphur dioxide from the volcanic Laki craters took their toll on the populace too, the governing Danish authorities suggested transferring the remaining Icelanders to Jutland for resettlement.

Climatic change was only one factor, although the most important one, in the decline of Iceland. The warmer period that began last century, ushered in a revival of the country. The history of Iceland illustrates that climatic change is nothing new and that mankind has had to face, and try to cope with its consquences. The Little Ice Age—that brought Iceland to penury—was a short-lived climatic change, a result of the cycle of warm and cold periods to which the world is subject.

GREENLAND

If Iceland survived climatic change, European settlement in medieval Greenland did not.

Norse settlement in Greenland began in 985 or 986, when Erik the Red led a fleet of over two dozen small ships from Iceland. Around 900 Gunnbjorn Ulfson, carried West in a storm, had chanced upon the skerries off the eastern shore of Greenland. In 978 Snaebjorn Galti and some companions, fleeing justice in Iceland, decided to find this new western land but harsh conditions forced their return after a long winter. In 981 or 982 Erik the Red followed their route and found land 290 km (180 miles) Northwest of Iceland. Initially he, too, discovered only icy wastes, but rounding Cape Farewell, Erik found a pleasant ice-free land with good natural pastures after which he named the country Greenland. The climate was mild enough for his party to winter there successfully and to encourage him to bring settlers back to this new frontier.

Within a couple of seasons about 500 Icelanders had settled in the area around pres-

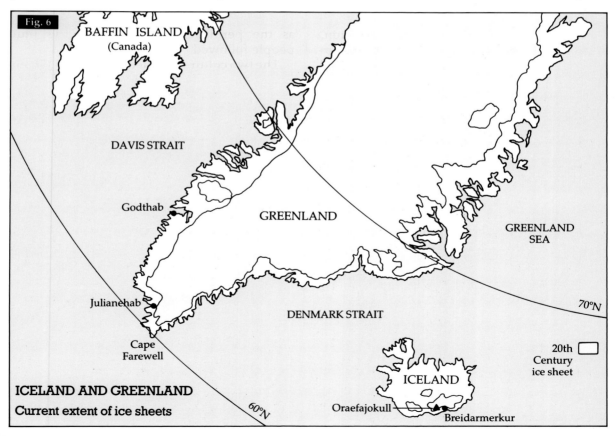

Fig. 6

BAFFIN ISLAND
(Canada)

DAVIS STRAIT

Godthab

GREENLAND

GREENLAND
SEA

Julianehab

70°N

Cape
Farewell

DENMARK STRAIT

20th
Century
ice sheet

ICELAND AND GREENLAND
Current extent of ice sheets

60°N

ICELAND

Oraefajokull

Breidarmerkur

ent-day Julianehab and before too long a similar community of sheep farmers was established around present-day Godthab (Nuuk).

For over 250 years an independent Norse republic flourished in Greenland. More than 3000 European people lived in early medieval Greenland, farming livestock on pastures that lined the edges of deep fjords that were rich in fish. Sheep, cattle, pigs and goats were all reared successfully and enough hay could be made to keep sufficient animals through the winter.

This thriving farming community was made possible by a climate that was probably milder than it is at present. The Greenlanders were virtually self-sufficient in food and maintained a healthy trade with northern Europe in walrus ivory, bear furs and seal products.

Greenland suffered double blows in the 13th century—one man-made, the other climatic. The loss of independence to Norway in 1261 damaged the economy. Henceforth, Norway maintained a legal monopoly of trade with the island, sending only two ships a year to Greenland to collect exports. After a while, visits by the Norwegian trading ships became less regular and gradually ceased altogether as Norway lost interest in her distant possession.

The greater blow, though, was climatic. During the 13th century temperatures decreased worldwide. Winters in Greenland grew more extreme and the growing period became too short to sustain agriculture. The deterioration of the weather in the island was much greater than it was in Iceland. Glaciers expanded, covering pastureland. Famine decimated the 12000 to 15000 Greenlanders whose links with the outside world became ever more difficult as pack ice spread down the coast. By the 15th century all routes to Greenland were practically impassable and famine became common. The climatic change seriously altered the conditions that had made European settlement of Greenland possible and then isolated the island in an impenetrable barrier of ice.

Cut off from Europe, the declining community was forced to attempt self-sufficiency. The Greenlanders were able to feed themselves still but increasingly poorly. They remained able to hunt and a reduced livestock population could be maintained on the diminishing pastures. Fish, too, continued to be available.

But a diet of meat and fish had drastic consequences on the health of the people. Deprived of trade—and therefore of fruit and grain—a weakened populace grew sickly. Deprived of imported wood, the Greenlanders could no longer build boats; deprived of iron, they could no longer make farming implements or arrows for hunting—nor could they make weapons for defence.

The warmer conditions that had allowed Icelanders to settle and farm in Greenland in the early medieval period had also encouraged the polar bear and walrus populations of the Arctic to move North. Greenland's Inuit people (Eskimos) followed the animals upon which their livelihood depended. In the period of the Norse republic in Greenland there was virtually no contact between the European settlers, who moved into an uninhabited land, and the Inuit population who had relocated further North. During global cooling in the 13th century, thick ice began to cover the northern coastal districts of Greenland as well as Baffin Land and many of the other islands of the Canadian Arctic. Polar bears migrated South across the ice that covered Davis Strait and Baffin Bay. Walruses were pushed South as the permanent icecap grew. The Inuit people followed.

The two cultures clashed in southern Greenland and the Greenlanders, enfeebled by disease and malnutrition, were unable to withstand the attacks they came under when the Inuit people tried to take over their land.

When the climate briefly ameliorated in the 16th century—and passage to Greenland was possible again—the European community was no more. The combination of disease, of famine resulting from climatic change and of attacks from the Inuit tribes, who had been forced South by the increasingly extreme conditions, had wiped out the Greenlanders. Only the deserted ruins of their buildings and small herds of cattle that had gone wild survived. Even the memory of the small distant Norse outpost had been almost forgotten. There was little incentive to resettle. The Spanish and Portuguese explorers were opening up a wider, warmer, and altogether richer New World. Whaling drew some interest, but the climate deteriorated again, reaching its nadir in the 18th century, when it was not unknown for polar bears from Greenland to reach Iceland

A small ice-bound fjord on the Greenland coast. When Greenland was first discovered by Europeans—Irish monks 1000 years ago—mosquitoes rather than ice were a problem. (The Hulton Deutsch Collection)

over the pack ice that covered the Denmark Strait.

The end of the Little Ice Age in the early 19th century enabled increased activity in Greenland. Hunting and fishing prospered and the Danish authorities did much to develop the island. This century the climate has been significantly warmer, and in the 1930s temperatures reached a maximum that may have approached the mild conditions that Greenland had enjoyed in the 11th and 12th centuries.

In response to warming, the seal population again moved North and there was a considerable increase in schools of fish in coastal waters. Yet by the 1970s and early 1980s temperatures in Greenland were falling again, a phenomenon that was used by some to suggest that global cooling was a potential danger. Indeed, during the 1970s temperature decreases had seriously diminished the stocks of cod in Greenland waters.

Holes drilled into the Greenland icecap by US Army engineers have enabled scientists to study the ice core to a depth of over 1400 m (4600 ft). This has revealed a pattern of climatic variation over the last 100 000 years and suggests an imminent cool period. It could be said that the temperature decreases of the early 1970s were the first signs of this cooling. Now, however, the temperatures recorded in Greenland are rising once more. This warming is usually ascribed to the greenhouse effect, but—as we shall see—other factors may be involved.

An understanding of what is happening was further complicated by the revelation in August 1990 that much of the Greenland icecap is not retreating as had been supposed (see p. 169).

Greenland and Iceland are examples of what can happen to societies when temperatures decrease and rainfall, or snowfall, increases. A climatic change of a different type was involved in the following examples from the Middle East and North America in which decreases in rainfall, and in one case an increase in temperature, were involved.

ROSE RED CITY

The road between the ancient Jordanian town of Kerak and the 'rose red city' of Petra suddenly dips and snakes into a deep, wide and impressive canyon at the foot of which trickles the insignificant little River Hasa. The landscape, which resembles a smaller version of the American Grand Canyon, is semi-desert, but the valley and many of its physical features are the result of normal river erosion far in excess of anything that could be accomplished by the diminutive Hasa. The conclusion is that this region must have been considerably wetter in the recent geological past. And from Petra to the South and Kerak to the North there is further evidence of a different climate in the more recent historical past.

According to the Bible, Moses struck a rock in the wilderness and water gushed forth for the Israelites. The place is still called the Valley of Moses—or in Arabic *Wadi Musa*—but the spring is tiny and unreliable. Nearby are the remains of the water channels and the runnels cut into the rock that brought supplies of water to the famous city of Petra. The former capital of the Nabateans, an Arab people who set up a kingdom in the region about 310 BC, Petra is cut out of red standstone cliffs and can only be approached through a narrow gorge called the Sik. Magnificent rock-cut temples and tombs remain from the Nabatean city, with ruined classical buildings constructed by the Romans who took Petra in AD 106. Today, Petra is dry and dusty, and looking around its vast natural amphitheatre, it is impossible to believe it was once a rich city with a flourishing agriculture. The desert sand at the foot of the temples was once small fields and gardens, carefully tended with irrigation water which was stored in deep tanks cut from the rock. Petra was a great trading city, a centre of the spice trade. The site was always marginal, near the edge of arid land, but there is very considerable evidence from the archaeological record that Petra was once more Mediterranean in climate, probably with a reasonable rainfall total in the winter to sustain it through the long hot summer.

At the beginning of the Islamic period in the 7th century AD, the city was abandoned. The usual reason given for the decline of Petra is a change in the trading routes, but if the city was still important would it not have still attracted

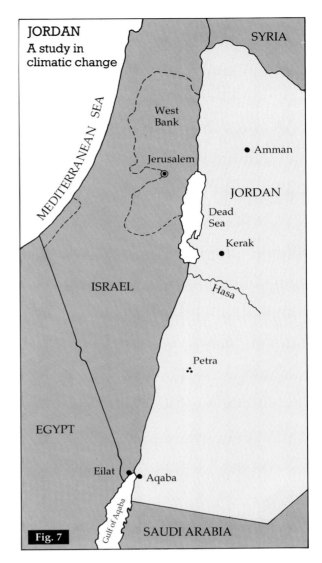

JORDAN
A study in climatic change

Fig. 7

commerce? An alternative factor in the death of Petra is gradual climatic change, a prolonged and increasing drought that dried up the ingeniously constructed water system that supplied the city.

All knowledge of Petra disappeared from Western history until the Swiss explorer Johann Ludwig Burckhardt stumbled upon the ruins in 1812. Engravings from the early 19th century revealed the beauty of the 'rose red city, half as old as time' to a European audience. One of the pictures portrays a flowing stream in front of the first temple at the end of the Sik; by the 1970s not a trace of that stream remained. Human activity cannot be the reason for its disappearance, as the area supported only a tiny number of wandering

Bedouin herdsmen. A gradual natural climatic change appears to have been at work.

Perched on a high butte, the Crusaders' castle of Kerak occupies a natural defensive site. Kerak was one of the ancient Biblical capitals of Moab. Later it was a Byzantine walled · city and the seat of an important bishopric. In 1110 French Crusaders occupied the area and created for themselves the principality of Outre-Jourdain ruled from the huge citadel they erected. The land around Kerak was desirable, a country of wheat, fruit, vegetables and pastures. Kerak was a major city in an economically viable region. Its collapse into a dusty region, supporting a sparse population by poor farming, cannot be explained by the dramatic overthrow of the Crusaders at the Battle of Hattin in 1188. The rulers may have changed, but the local people remained and their agriculture slowly declined as Kerak became drier over the centuries.

In southern Jordan we have an historical record of a gradual decline in the lifestyle and the farming of an area that is usually ascribed to history and changes of rulers and the fall of ancient empires. But that degradation has been a gradual process over nearly 2000 years and, though it can be partly understood through historical events, it is at the same time a remarkable record of climatic change.

HIGH TIDE

Few changes in nature are as dramatic as a change in sea level and one of the possible consequences of global warming, owing to the greenhouse effect, may be a rise in sea level (see p. 128–140). Although the waves are constantly forming and breaking, the oceans themselves seem to be more enduring. But virtually all the land surfaces of the world were at one period of geological time or another beneath the sea. Even in a short period of time the extent of the seas can change, often suddenly as when the North Sea broke into the Netherlands to create the Zuyder Zee (see p. 135).

But have there been changes of sea level as a result of climatic change in recent history? If so, a study of the past could give an idea of

Left Kerak was the largest of a line of Crusader castles stretching from Turkey in the North to Aqaba in the South. Although the castle is in ruins, what remains is vast in scale. Steep natural defences protected Kerak on three sides but on the western side was a moat, which is now bone dry. The climate in the region during the Middle Ages must have been appreciably milder to have provided the water to fill the moat. (Spectrum Colour Library)

Right Fishing boats at Rye in Sussex, moored close to the site of the original harbour. As late as the 18th century there was still an anchorage almost outside the town walls. (Spectrum Colour Library)

GRAPES OF WRATH

The rolling agricultural landscape of the US state of Oklahoma was the setting for one of the best known climatic variations of the 20th century. Western Oklahoma is the part of the state with the lowest rainfall. Between 400 and 500 mm (16 to 20 in) is the average. The region grows wheat and sorghum where the rainfall allows, while the drier areas support pasture for cattle ranching.

As with most marginal areas, the rainfall in western Oklahoma tends to be unreliable and in the 1930s it failed for several years. The soil became parched, cracked and turned to dust partly through lack of rain, and partly through over-intensive cultivation of land that was naturally better suited for grazing than the plough—thus a meteorological drought and an agricultural drought (see p. 83) were intertwined. In addition, dust clouds blocked out the light of the Sun by day and

thousands of farmers were ruined. Unable to retain their land, because of the prevailing system of agricultural economics, many 'Okies' were dispossessed. The story of one such family of Okies and their journey from the Dust Bowl of Oklahoma to California was immortalized by John Steinbeck in his novel *The Grapes of Wrath*.

At the time the prolonged drought was interpreted as a climatic change by some observers. Yet, in the 1940s the normal rainfall totals were re-established and the Dust Bowl was gradually reclaimed, although as the topsoil had been removed by wind erosion, different agricultural practices had to be employed in the impoverished region. The 1930s drought correlates to drought and warming in a number of other parts of the world—and might be regarded as evidence suggesting a natural climatic cycle.

During the 1930s, drought and overcultivation resulted in a large-scale removal of the topsoil of much of Oklahoma, west Kansas and northern Texas. This created what became known as the Dust Bowl. Many photographs, such as this one, were taken to capture the clouds of dust raised by the wind, but the photojournalist Dorothea Lange focused world attention on the human side of the disaster. (Explorer FPG)

what we might expect in the future if global warming were to result in a rise of a third of a metre or more before the middle of the 21st century, or even more later.

During the Little Optimum of the medieval period (see p. 60), temperatures were significantly above those of the present day. It is, therefore, likely that the icecaps were less extensive then and the sea level consequently higher. For possible evidence of this we should look for changes in the configuration of the coastline since the Middle Ages, and such a difference can indeed be found on old maps of the low-lying marshy lowlands where Kent meets Sussex.

Today, the charming hill top borough of Rye sits proudly clustered around the sturdy tower of a fine parish church, gazing out across two miles of pancake flat fields to the English Channel. To the South, the gridiron plan of the medieval 'new town' of Winchelsea resembles a small 'urban village' in parkland, but a hint of its former importance can be gained from its fine gates. On the edge of the town the ground falls away sharply to the levels, flat ground stretching to the sea over a mile away. The two towns, each on a hill, face one another across marshy ground that was once sea. Both were major ports, members of the confederacy of the Cinque Ports, an alliance of small ports along the South coast that was granted privileges by the Crown in return for raising ships and men to form the nucleus of a navy in times of war. Both towns flourished through trade and piracy, but during the later part of the Middle Ages the estuaries on which they were set silted up. By 1568 the topographer William Camden could write of Rye that 'it beginneth to complain that the sea abandonneth it'.

Such a change is consistent with a fall in sea level as temperatures fell with the approach of the Little Ice Age. We know that the icecaps became greater in extent from the 14th century and there is evidence for resultant changes in sea level. The unusual explanation given for the demise of the ports of Rye and Winchelsea is silting of the channels, but these ports had existed for several hundred years without widescale dredging. The result of climatic change at the beginning of the Little Ice Age, is a very likely culprit for the sudden onset of silting and a fall in the sea level, although other local factors also played an important role.

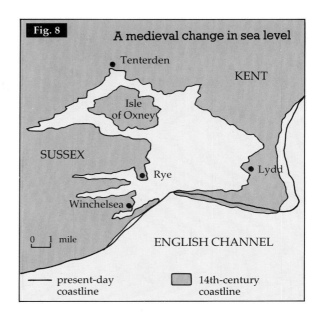

Fig. 8 — A medieval change in sea level

The small town of Winchelsea illustrates the change graphically. Old Winchelsea was built on a low flat island in the combined estuaries of the Rivers Rother and Brede. By the early 13th century—when sea levels were higher—it was a prosperous town, the most important of the Cinque Ports, earning a good living exporting wool. But in 1250 things began to go wrong for Winchelsea. The 1250s, when the climate was beginning to slide into the Little Ice Age, saw storms of an unprecedented ferocity. In 1250 over 300 houses were washed away by the sea and in 1252 further damage was done. In 1287 a storm—which from the accounts of the time sounds much like the 'hurricane' that hit Southeast England in October 1987—completed the destruction and a combination of high winds and high waves wiped out most of what remained of Winchelsea.

A decision was taken to rebuild Winchelsea as a planned medieval town, but the sea started to retreat as the sea level fell. The port was never completed and the grand wharves and warehouses were never constructed as repeated dredging could not keep the anchorages clear.

The accompanying map shows the extent of the retreat of the sea, associated, in part, by a change of sea level. So great were the changes in this small region that Tenterden, now a market town almost 18 km (10 miles) inland, was in the 12th and 13th centuries a small but flourishing seaport.

6 Climatic controls and effects

What is happening to our weather? The question should be 'What is happening to our *climate?*' for the weather is *always* changing.

The terms 'weather' and 'climate' are often misused and even interchanged in everyday speech. Weather is what we experience on a day-to-day basis in terms of temperature, rainfall, sunshine, cloud cover and so on. Tuesday may be overcast and humid, while Wednesday may be hot and sunny with clear skies. By its very nature, weather is changeable unless you happen to live, for example, in some equatorial places where on the majority of days the morning is warm and sunny, a thick cloud cover may form by noon and a heavy rain storm may occur each evening. But in most of the world the weather is not so predictable—if it was, there would be no need for a weather forecast.

If, in much of the world, daily variations in weather are to be expected, should we not expect some consistency in the climate? Climate has been described as 'average weather', usually over a period of years.

The long-term weather conditions in any area reveal enough similarity or enough of a pattern to be recognized as a type, in other words a particular kind of climate. We can, for example, delineate a broad Mediterranean type of climate, with hot dry summers and warm wetter winters; or a warm temperate climate such as that experienced in Britain, with moderate temperatures, which reach a maximum in midsummer, and rainfall at all seasons.

Unfortunately, the definition of different climates is not simple. Although about a dozen broad types are generally recognized, there are many regional and even local variations. One study of climate in Spain acknowledged well over 100 different local climates or microclimates.

The type of climate experienced in any given area depends upon a combination of many factors including how far the area is from the Equator or the poles (latitude), how high or low-lying the area is (altitude) and how far inland it is situated (continentality).

Recording rainfall in the 19th century. The Rev. Griffiths, a keen amateur meteorologist, established 42 experimental rain gauges in his garden at Strathfield Turgiss rectory in Hampshire, England.

SOLAR RADIATION

The climate of any given area is a result of various factors, the most important of which is the amount of incoming solar radiation and the amount of outgoing radiation from the Earth. This, in turn, depends upon the latitude of the place concerned and upon the season.

Because the Earth's axis is inclined, the Sun's radiation is not received to an equal degree throughout the planet. Thus on 21 June, the North Pole is tilted towards the Sun and places in the northern hemisphere receive their maximum amount of radiation, while locations in the southern hemisphere, which are tilted away from the Sun, receive the least solar radiation. This is the real Midsummer Day, the summer solstice.

The reverse situation is true on 21 December, when the South Pole is tilted towards the Sun and the northern hemisphere experiences Midwinter Day. The middle position occurs on 21 March and 21 September (the equinoxes) when all places in the world experience the same length of day and the Sun's rays are directly overhead at local noon at the Equator.

The axis of rotation of the Earth is inclined at an angle of 23.5 degrees from the vertical. The seasons are caused by the annual revolution of the planet in an elliptical orbit round the Sun, and the daily rotation of the Earth upon this inclined axis. The Sun's rays, therefore, strike the Earth at varying angles (see Figure 9).

In the diagram, it can be seen that in summer in the northern hemisphere, the Sun's rays reach the surface of the planet more directly through the atmosphere. But in winter the Sun's rays reach the northern hemisphere obliquely and so give less heat.

If the temperature of a location depended just upon the inclination of the globe, the warmest place on Earth on 21 June would be the North Pole. However, the *angle* of incidence of solar radiation determines the *effectiveness* of that energy. When the Sun is directly overhead at any given spot there is maximum heating as the solar energy is absorbed by the Earth. Yet on 21 June, the great amount of solar radiation received at the North Pole is spread out over a wide area and this is, therefore, far less effective than the more concentrated heat received in the tropics.

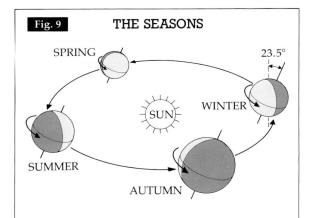

Fig. 9 **THE SEASONS**

SPRING — 23.5° — WINTER — SUN — SUMMER — AUTUMN

The passage of the seasons in the northern hemisphere is shown in this diagram. The tilt of 23.5° of the Earth's axis allows solar energy to reach the surface of the globe more directly through the atmosphere. In winter the reverse situation applies.

The nature of the Earth's surface at any given spot is also a powerful climatic determinant. The specific heat of land and of the sea differ sharply. Any given amount of solar energy will increase the temperature of the continents more quickly than it will increase the temperature of the oceans. Thus the land is warmer than the sea by day and in summer. As the land gives out heat more rapidly than water, the sea is warmer than the land by night and in winter.

The third main determinant of the climate of any place is atmospheric pressure. Cool air is denser than warm air, so cool air tends to form areas of high pressure at the Earth's surface, while warm air tends to form areas of low pressure. Thus, atmospheric pressure is uneven across the face of the globe and air moves—as wind—from regions where the pressure is high to places where the pressure is low in an attempt to rectify this inequality.

There is an overall movement of air—a wind system—from the Equator to the Poles, but this is complicated by the rotation of the Earth which deflects the wind from a straight path to the right in the northern hemisphere and to the left in the southern hemisphere. This deflection is known as the Coriolis effect.

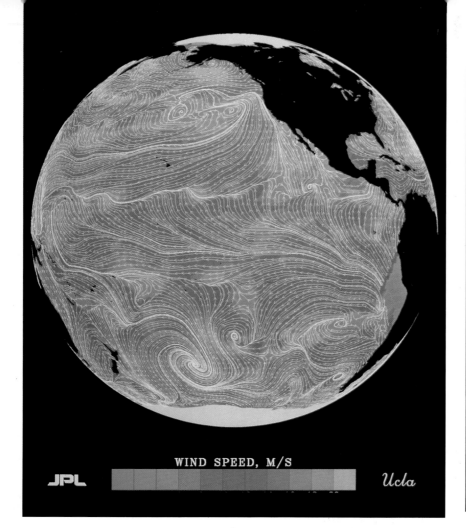

WIND SPEED, M/S

JPL   Ucla

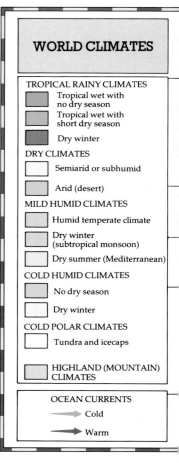

WORLD CLIMATES

TROPICAL RAINY CLIMATES
  Tropical wet with
  no dry season
  Tropical wet with
  short dry season
  Dry winter
DRY CLIMATES
  Semiarid or subhumid
  Arid (desert)
MILD HUMID CLIMATES
  Humid temperate climate
  Dry winter
  (subtropical monsoon)
  Dry summer (Mediterranean)
COLD HUMID CLIMATES
  No dry season
  Dry winter
COLD POLAR CLIMATES
  Tundra and icecaps

  HIGHLAND (MOUNTAIN)
  CLIMATES

OCEAN CURRENTS
  → Cold
  → Warm

Above An enhanced satellite photograph of the Earth's wind speed patterns. A dominant East–West pattern is visible (see p. 79). The Coriolis effect would lead one to expect a North–South pattern. (Telegraph Colour Library)

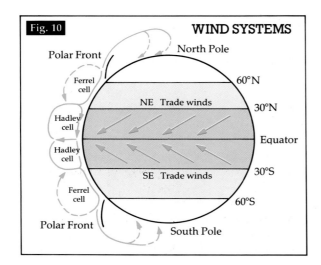

Fig. 10   WIND SYSTEMS

Polar Front   North Pole
Ferrel
cell   60°N
Hadley   NE Trade winds   30°N
cell
Hadley   Equator
cell
Ferrel   SE Trade winds   30°S
cell   60°S
Polar Front   South Pole

The atmosphere has been called a heat machine in which the wind systems originate in a tropical heat source and flow to a polar sink, but when the complication of other major zones of high and low pressure are added to the picture a simplified world wind system consists of six cells in which the air flows at upper and lower levels (see Figure 10).

Immediately North and South of the Equator are the Trade wind cells (also known as the Hadley cells) in which warm air rises at the Equator and flows away from the Equator at upper levels, to sink at the Tropics. Winds blow from the Tropics back to the Equator and towards the Poles. The latter flow is part of the midlatitude cell (or Ferrell cell) in which air masses rise at the polar fronts in latitudes of 50 to 60 degrees. This rising warm air divides between a weaker return flow at upper levels—which is part of the midlatitude cell—and a stronger upper flow towards the Poles—which is part of the Polar cell. Air masses sink in high latitudes and flow, as cold air, away

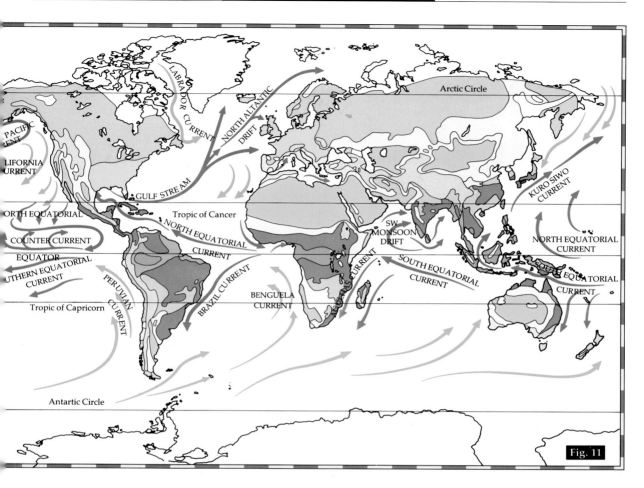

Fig. 11

from the Poles and eventually form parts of the Polar front. Yet, glancing at a map of the wind systems of the world, there appears to be no obvious North–South alignment. The majority of air movements appear as East–West or West–East owing to the substantial modification of the basic wind cells, largely because of the Coriolis effect.

The distribution of land and sea, and of high and low ground, also determine climate. The world's climatic belts do not exhibit a regular and gradual decrease in temperature from the Equator to the Poles. Mountain ranges may block or divert the movement of air masses, or change their nature as they rise and cool over their heights. Ocean currents may warm or cool areas to a degree that is atypical of places in their latitude, and distance from the sea also plays a part in modifying the climate.

When air masses 'stagnate' over land or sea, the air acquires the characteristics of the area. In this way air that remains for any length of time over the Arabian peninsula becomes hot and dry, while air that remains over the adjoining Arabian Sea becomes moist and warm. Thus air masses take their characteristics from both the region of their origin and the areas they pass over on their path. For example, cold air from over Canada moves East and is modified, becoming warmer and moister, as it crosses the Atlantic on its way to Northwest Europe. The air starts out as Arctic continental, but by the time it reaches Britain it is transformed into Polar maritime.

CLASSIFICATION OF CLIMATE

The climate of each and every location is unique. No other place combines the same elements of latitude, aspect, height above sea level, continentality (that is distance from the sea), exposure and so on. Such a localized experience of climate is known as a microclimate.

On the larger scale, climate is classified into

broad types in a variety of ways. The ancient Greeks were the first to attempt to define and classify climate.They recognized a winterless tropical zone in Africa and the Middle East, a summerless polar zone in the far North of Europe and Asia and an intermediate zone—which we now know as the temperate zone—in which there are cool summers and mild winters.

Most systems of climatic zones, and there are many, are based upon two elements of weather—precipitation in its various forms (rain, snow, hail, etc.) and temperatures. Thus each location is grouped with places experiencing similar totals and seasonal distributions of rainfall and temperature. In this way, climatic zones such as the mild humid climate with a dry winter can be defined (see Figure 89).

Temperature and precipitation are two of the climatic effects. Others include wind and humidity. The climatic effects are the result of climatic controls which include ocean currents, altitude and continentality.

Distance from the sea (continentality) is a major influence on the climate of an area. As the water heats up more slowly and cools more gradually than the land, coastal places in the temperate (middle) latitudes have cooler summers and milder winters than places that are further inland. Thus Ireland is said to have a maritime climate while Poland, in much the same latitude, is said to have a more continental climate—westerly air masses reaching Poland have already crossed France, the Low Countries and Germany and have been modified on their journey.

Maritime climates are influenced not only by the proximity of the sea, and by its moderating influence, but also by the currents flowing in those waters. Cool ocean currents can give a place a considerably colder weather than its latitude would suggest, for example, Newfoundland is cooled by the cold Labrador Current which flows South through Davis Strait between Baffin Island and Greenland. In the same latitudes, the British Isles have substantially higher temperatures because of the

The Quito Valley in Ecuador where—because of an altitude of 2860 m (9400 ft)—the climate is perpetually 'spring-like' despite the location almost on the Equator. (Gamma)

effects of the warm North Atlantic Drift, a powerful current that is the continuation of the Gulf Stream which flows from the Gulf of Mexico.

Relief, too, is a powerful determinant of climate. Temperature falls or lapses with altitude, hence the 'spring-like' climate of Quito in Ecuador which is almost exactly on the Equator but which is far cooler than its latitude would suggest. To journey between the two principal cities of Ecuador is not only to climb many thousands of feet but also to experience a change of climate that is like going from one season to another. Guayaquil, on the Pacific Coast, experiences a typical tropical maritime climate with average temperatures between 23°C and 26°C (73°F and 79°F), while Quito, barely 400 km (250 miles) away, enjoys average temperatures of 12°C to 18°C (54°F to 64°F).

As well as being cooler, mountain ranges are often rainier than adjoining lowlands. Warm moist air rises over the mountains and cools. It condenses and thus produces either rain or snow. In most areas of the world, there is a prevailing direction of the wind, for example the westerlies over Ireland. If rain-bearing winds predominate in uplands from one particular direction, the land on that side will be wetter than the much drier area in the rain shadow on the other side. In South America, for example, the mountains of Chile experience a very high rainfall, while the semi-desert of Patagonia lies in the rain shadow on the eastern side of the Andes, and experiences low rainfall.

RAIN AND DROUGHT

A useful distinction has been made between two types of drought—meteorological drought, which is a fall in rainfall over a measured period, and agricultural drought, which is a fall in the moisture content of the soil. The former may not necessarily have much impact on people's lives and livelihood; the latter will as it leads to a fall in agricultural production. A meteorological drought can be identified with ease by the record of lower precipitation. An agricultural drought is recorded in terms of reduced crop yields, and, if prolonged, the death of livestock and famine, and although the two are obviously linked, a meteorological

drought is not the sole cause of an agricultural drought. Deforestation, farming practices and other human activities may combine to produce an agricultural drought, the result of which, in terms of erosion, landscape and flora, may help trigger or increase a meteorological drought.

It has been suggested that overgrazing in the Sahel has helped to prolong a meteorological drought (see p. 14). Overstocking the poor natural grasslands in countries such as Mali and Burkina Faso degrades the pasture, thinning the grass cover. It has been claimed that this sparser cover of grass is more effective at reflecting solar radiation. In this way, it is claimed, less solar energy is absorbed by the surface of the Sahel. The atmosphere becomes more stable and less rainfall results. This theory proposes a cumulative effect with ever decreasing absorption of solar radiation, ever greater reflection, progressively less rainfall, and progressively greater degeneration of the grasslands.

The agricultural drought could be contributing to the meteorological drought in another way as well. It has been suggested that the large amounts of airborne dust, in part a result of farming practices, are actually preventing the formation of raindrops.

Raindrops are formed at the end of a chain which begins with the lifting of damp air from the land or sea surface. As it rises, the damp air expands—owing to the lower air pressures into which it moves—and in expanding, it cools. In cooling, the relative humidity increases and the air becomes saturated with water vapour, which condenses around minute airborne particles such as sea salt and dust. By collision and attraction, these droplets grow and coalesce within clouds. The droplets will be of differing sizes as the particles around which they formed will vary in dimensions. The larger droplets will move rapidly within the cloud, collide with small ones and grow by accretion (see Figure 12).

Outside the tropics, most rain starts its journey from a cloud as ice. Clouds are maintained by air currents from below. Eventually a cloud, or a part of a cloud, will be raised to a height at which it contains much supercooled water below 0°C (32°F). Some of the droplets may freeze and can no longer be held by the cloud. Others may grow through colliding with other

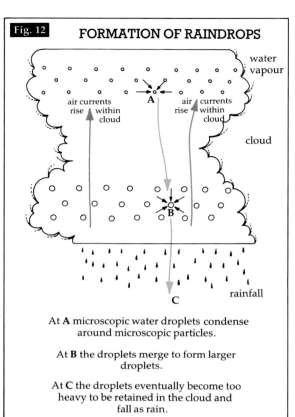

FORMATION OF RAINDROPS

Fig. 12

water vapour

air currents rise within cloud

air currents rise within cloud

cloud

A

B

rainfall

C

At **A** microscopic water droplets condense around microscopic particles.

At **B** the droplets merge to form larger droplets.

At **C** the droplets eventually become too heavy to be retained in the cloud and fall as rain.

Above The rigid structure of a snowflake is due to hydrogen bonding. Only a six-branched shape is stable, but despite their essential similarity, no two snowflakes are identical. (Science Photo Library)

Left Killarney Lakes, Ireland. In the mild, wet maritime climate of Ireland skies are often overcast, with rainfall on two days out of three. (Spectrum Colour Library)

liquid or frozen droplets on their descent. In this way raindrops are formed. (Showers in tropical regions may be produced by clouds whose higher reaches do not reach the critical low temperature of 0°C—when larger droplets coalesce within the turbulent clouds to such an extent that they form drops large enough to fall as rain.)

As particles are required for condensation to occur, so that droplets can form in the first place, the superabundance of minute dust particles, produced from eroding soils, would on the face of it seem to be a good thing. With so many particles available, surely condensation would occur, clouds would form and eventually rain would be assured? There is, however, an (as yet undefined) critical number and size

of particles. If farming practices are producing too much dust in the air, too little water will condense around each individual particle and droplets of insufficient size will grow for raindrops eventually to form. Thus farming practice can help to create drought.

CLIMATIC BELTS

The climatic controls and effects outlined above shape the world's many different climates. Any classification can recognise only the very broadest climatic types, the borders between which may be wide zones of gradual change.

Tropical grassland near Nairobi in Kenya, a classic savannah climate or tropical rainy (dry winter) climate. Defining climatic types, or climatic regions, is of varying usefulness, and only parts of the world where conditions are either a constant or predictable are climatic 'types' a reliable guide to prevailing weather.

Most climatic divisions of the world derive from the classification of Vladimir Köppen (published in 1900) which is based on a correlation of temperature and rainfall averages with vegetation. (Spectrum Colour Library)

Tropical rainy climates

The areas marked on the map on p. 81 as experiencing tropical rainy climates are more traditionally divided into three types—the equatorial climates, the tropical climates and the monsoon climates. To a greater extent the equatorial climates correspond with the belt marked as experiencing a tropical wet climate with no dry season. These areas lie within the zone of convergence of the Northeast and Southeast trade winds (see Figure 10). They are regions of great uplift of moist air, and therefore of heavy rainfall, and they gain their high temperatures from their position near to the Equator. The equatorial belt is one of considerable stability and of little to no seasonal variation.

On either side of the equatorial climatic belt, and in South America occupying a greater area than the equatorial zone, is a region of tropical climates which experience a short dry season. In some cases there is no pronounced dry season but rather two rainfall maxima—for example in parts of East Africa. These tropical climates owe their more seasonal nature to the fact that, for part of the year, they lie within the zone of convergence of the trade winds, while for the rest of the year they are under the influence of those winds.

The tropical zone which experiences a dry winter is subject to the trade winds to a greater extent. Inland and on the western side of the land masses, the dry period is prolonged—this is when dry air is blowing from the continental interiors. The tropical monsoon zone of South Asia lies within this belt, marked on the map on p. 81. The monsoon is caused by the diversion of the Southeast trades from South of the Equator towards India, drawn in by the strength of a deep low pressure area in Asia in the summer.

Tropical rainy weather stations—Temperature

Station	Average temperature in °C (*Fahrenheit in italics*)											
	J	F	M	A	M	J	J	A	S	O	N	D
Bombay, India	24	24	27	28	30	29	27	27	27	28	27	25
	76	*76*	*80*	*83*	*86*	*84*	*81*	*81*	*81*	*82*	*81*	*77*
Calcutta, India	19	22	27	29	30	29	29	28	28	27	23	19
	67	*71*	*80*	*85*	*86*	*85*	*84*	*83*	*83*	*81*	*73*	*67*
Manila, Philippines	25	26	27	28	28	28	27	27	27	27	26	25
	77	*78*	*80*	*83*	*83*	*82*	*81*	*81*	*80*	*80*	*78*	*77*
Rangoon, Myanmar	25	26	29	31	29	27	27	27	27	28	27	25
	77	*79*	*84*	*87*	*84*	*81*	*80*	*80*	*81*	*82*	*80*	*77*
Caracas, Venezuela	18	18	19	20	21	21	20	20	21	20	19	18
	65	*65*	*66*	*68*	*70*	*69*	*68*	*68*	*69*	*68*	*67*	*65*
Banjul, The Gambia	23	24	24	24	25	27	27	26	27	27	26	26
	74	*75*	*76*	*76*	*77*	*80*	*80*	*79*	*80*	*81*	*79*	*75*
Colon, Panama	27	27	27	27	27	27	27	27	27	27	27	27
	80	*80*	*80*	*81*	*81*	*80*	*80*	*80*	*80*	*80*	*80*	*79*
Jakarta, Indonesia	26	26	26	27	27	26	26	26	27	27	26	26
	79	*79*	*80*	*81*	*81*	*80*	*80*	*80*	*81*	*81*	*79*	*79*
Nairobi, Kenya	18	18	19	18	17	16	15	16	17	19	18	17
	64	*65*	*66*	*65*	*63*	*61*	*59*	*60*	*63*	*66*	*64*	*63*
Manaos, Brazil	27	27	27	27	27	27	27	28	28	28	28	26
	80	*80*	*80*	*80*	*80*	*80*	*80*	*81*	*82*	*83*	*83*	*82*
Yaoundé, Cameroon	23	23	23	22	22	22	21	22	22	22	22	23
	74	*74*	*74*	*72*	*72*	*71*	*70*	*71*	*71*	*71*	*72*	*73*
Kananga, Zaire	24	24	24	25	25	24	25	24	24	24	24	25
	76	*76*	*76*	*77*	*77*	*76*	*77*	*76*	*76*	*76*	*76*	*77*

Differences in conversion are due to the figures quoted being originally in Fahrenheit.

Tropical rainy weather stations—Rainfall

Station	Average rainfall in mm (*inches in italics*)											
	J	F	M	A	M	J	J	A	S	O	N	D
Bombay	3	3	0	0	18	251	610	368	269	48	10	0
	0.1	*0.1*	*0*	*0*	*0.7*	*19.9*	*24*	*14.5*	*10.6*	*1.9*	*0.4*	*0*
Calcutta	10	25	36	56	127	302	323	340	254	124	15	5
	0.4	*1*	*1.4*	*2.2*	*5.6*	*11.9*	*12.7*	*13.4*	*10*	*4.9*	*0.6*	*0.2*
Manila	20	10	20	33	114	234	439	406	363	170	132	79
	0.8	*0.4*	*0.8*	*1.3*	*4.5*	*9.2*	*17.3*	*16*	*14.3*	*6.7*	*5.2*	*3.1*
Rangoon	5	5	8	36	307	467	546	432	391	185	71	8
	0.2	*0.2*	*0.3*	*1.4*	*12.1*	*18.4*	*21.5*	*19.7*	*15.4*	*7.3*	*2.8*	*0.3*
Caracas	23	8	15	41	71	102	109	107	104	97	84	46
	0.9	*0.3*	*0.6*	*1.6*	*2.8*	*4*	*4.3*	*4.2*	*4.1*	*3.8*	*3.3*	*1.8*
Banjul	0	0	0	0	5	74	277	498	254	94	5	3
	0	*0*	*0*	*0*	*0.2*	*2.9*	*10.9*	*19.6*	*10*	*3.7*	*0.2*	*0.1*
Colon	94	41	41	109	315	338	406	376	318	384	526	290
	3.7	*1.6*	*1.6*	*4.3*	*12.4*	*13.3*	*16*	*14.8*	*12.5*	*15.1*	*20.7*	*11.4*
Jakarta	330	325	198	130	102	94	66	43	74	114	140	216
	13	*12.8*	*7.8*	*5.1*	*4*	*3.7*	*2.6*	*1.7*	*2.9*	*4.5*	*5.5*	*8.5*
Nairobi	48	107	76	211	132	51	20	23	23	51	147	89
	1.9	*4.2*	*3.7*	*8.3*	*5.2*	*2*	*0.8*	*0.9*	*0.9*	*2*	*5.8*	*3.5*
Manaos	234	229	244	216	178	91	56	36	51	104	140	196
	9.2	*9*	*9.6*	*8.5*	*7*	*3.6*	*2.2*	*1.4*	*2*	*4.1*	*5.5*	*7.7*
Yaoundé	41	69	150	231	206	114	66	84	193	226	150	50
	1.6	*2.7*	*5.9*	*9.1*	*8.1*	*4.5*	*2.6*	*3.3*	*7.6*	*8.9*	*5.9*	*2*
Kanaga	183	137	201	155	79	5	3	64	165	168	231	168
	7.2	*5.4*	*7.9*	*6.1*	*3.1*	*0.2*	*0.1*	*2.5*	*6.5*	*6.6*	*9.1*	*6.6*

Left 'Mediterranean' regions experience a relatively uniform dry tropical climate in the summer months and a changeable temperate wetter climate during the winter months. Much of the rainfall comes in thunderstorms as in this storm over Cyprus. (H. Herrington)

Above right Upward-moving air currents extend the top of a cumulonimbus cloud into the characteristic anvil shape, known in North America as a thunderhead. Centres of electrical charge are established, resulting in thunder and lightning. (Telegraph Colour Library)

Mild humid climates

There is very considerable variation in the different classifications of these climates. The principal aspect used in classifying the three broad types shown on the map, p. 81, is the seasonal incidence of rainfall.

The mild humid areas in upper latitudes, such as the Northwest of Europe and the South Island of New Zealand, are largely subject to the regular passage of westerly cyclones which bring a changeable pattern of weather (see p. 24). The more transitional areas—such as the southern states of the USA and the South of China—lie between the trade winds and the zone of cyclones. They are warmer and in the case of the southern Chinese coast experience a monsoon.

The temperate climate on the western margins—the Mediterranean climatic type—has many variations but all are united in experiencing a hot dry summer during which the prevailing winds blow from over the land and a milder winter during which cyclones bring rainfall.

Mild humid climate weather stations—Temperature

Station	Average temperature in °C (*Fahrenheit in italics*)											
	J	F	M	A	M	J	J	A	S	O	N	D
Lisbon,	11	11	12	14	16	19	21	22	20	17	14	11
Portugal	*51*	*52*	*54*	*58*	*60*	*67*	*70*	*71*	*68*	*62*	*57*	*52*
Paris, France	3	4	6	9	13	17	18	18	14	10	6	3
	37	*39*	*43*	*49*	*56*	*62*	*65*	*64*	*58*	*50*	*43*	*38*
Athens, Greece	9	9	11	15	19	23	27	27	23	19	14	11
	48	*49*	*52*	*59*	*66*	*74*	*80*	*80*	*73*	*66*	*57*	*52*
San Francisco,	9	11	12	12	13	14	14	14	16	15	13	11
USA	*49*	*51*	*53*	*54*	*56*	*57*	*57*	*58*	*60*	*59*	*56*	*51*
Cape Town,	21	21	20	17	15	13	13	13	14	15	17	20
South Africa	*70*	*70*	*68*	*63*	*59*	*56*	*55*	*56*	*58*	*61*	*64*	*68*
Washington,	1	2	6	12	18	22	25	23	20	14	8	2
DC, USA	*34*	*35*	*43*	*54*	*64*	*72*	*77*	*74*	*68*	*57*	*46*	*36*
New Orleans,	12	14	17	21	24	27	28	27	26	21	16	13
USA	*54*	*57*	*63*	*69*	*75*	*80*	*82*	*81*	*78*	*69*	*61*	*55*
Adelaide,	23	23	21	18	14	12	11	12	14	17	19	22
Australia	*74*	*74*	*70*	*64*	*58*	*54*	*52*	*54*	*57*	*62*	*67*	*71*
Dunedin,	14	14	13	11	8	7	6	7	9	11	12	13
New Zealand	*58*	*58*	*55*	*52*	*47*	*44*	*42*	*44*	*48*	*51*	*53*	*56*
Chungking,	9	10	14	20	23	27	28	30	25	20	15	10
China	*48*	*50*	*58*	*68*	*74*	*80*	*83*	*86*	*77*	*68*	*59*	*50*

Differences in conversion are due to the figures quoted being originally in Fahrenheit.

Mild humid climate weather stations—Rainfall

Station	Average rainfall in mm (*inches in italics*)											
	J	F	M	A	M	J	J	A	S	O	N	D
Lisbon	91	89	86	66	51	20	5	5	36	84	109	104
	3.6	*3.5*	*3.4*	*2.6*	*2*	*0.8*	*0.2*	*0.2*	*1.4*	*3.3*	*4.3*	*4.1*
Paris	38	36	41	43	48	53	56	53	48	58	48	51
	1.5	*1.4*	*1.6*	*1.7*	*1.9*	*2.1*	*2.2*	*2.1*	*1.9*	*2.3*	*1.9*	*2*
Athens	51	43	30	23	20	18	8	13	15	41	66	66
	2	*1.7*	*1.2*	*0.9*	*0.8*	*0.7*	*0.3*	*0.5*	*0.6*	*1.6*	*2.6*	*2.6*
San Francisco	122	91	79	25	18	3	0	0	8	25	61	117
	4.8	*3.6*	*3.1*	*1*	*0.7*	*0.1*	*0*	*0*	*0.3*	*1*	*2.4*	*4.6*
Cape Town	18	15	23	48	97	114	94	86	58	41	28	20
	0.7	*0.6*	*0.9*	*1.9*	*3.8*	*4.5*	*3.7*	*3.4*	*2.3*	*1.6*	*1.1*	*0.8*
Washington DC	81	76	89	84	91	99	112	102	79	79	64	79
	3.2	*3*	*3.5*	*3.3*	*3.6*	*3.9*	*4.4*	*4*	*3.1*	*3.1*	*2.5*	*3.1*
New Orleans	114	109	117	114	104	137	165	145	114	81	97	114
	4.5	*4.3*	*4.6*	*4.5*	*4.1*	*5.4*	*6.5*	*5.7*	*4.5*	*3.2*	*3.8*	*4.5*
Adelaide	18	18	25	45	71	79	69	64	51	43	30	25
	0.7	*0.7*	*0.1*	*1.8*	*2.8*	*3.1*	*2.7*	*2.5*	*2*	*1.7*	*1.2*	*1*
Dunedin	86	69	76	69	81	81	76	79	71	76	84	89
	3.4	*2.7*	*3*	*2.7*	*3.2*	*3.2*	*3*	*3.1*	*2.8*	*3*	*3.3*	*3.5*
Chungking	18	23	33	102	135	170	135	112	147	117	51	23
	0.7	*0.9*	*1.3*	*4*	*5.3*	*6.7*	*5.3*	*4.4*	*5.8*	*4.6*	*2*	*0.9*

Cold humid climates

The map on p. 81 indicates two main types of cold humid climate. Those with no dry season are under the influence of moist westerly winds and lie on the western side of continents. Those with a distinct dry season lie on the eastern side of the vast Eurasian continent. This eastern type—found in the Far East of the USSR and parts of northern China—is more continental. The western type—found across much of North America, in northern Sweden, Finland, and across the USSR—is more maritime. There is a gradual eastward modification of the climate to a greater annual and daily range of temperature, a lower rainfall and a marked summer rainfall maximum.

The continents of the southern hemisphere do not penetrate far enough South to experience these cold humid climates

Cereal cultivation in the subhumid climatic region in New South Wales, Australia. Both the humid temperate areas and the subhumid regions are important for cereal growing. (British Petroleum Company)

Cold humid climate weather stations—Temperature

Station	Average temperature in °C (*Fahrenheit in italics*)											
	J	F	M	A	M	J	J	A	S	O	N	D
Dawson, Yukon,	−31	−24	−16	−2	8	14	15	12	6	−4	−17	−25
Canada	*−23*	*−11*	*4*	*29*	*46*	*57*	*59*	*54*	*42*	*25*	*1*	*−13*
Winnipeg,	−20	−18	−9	3	11	17	19	18	12	5	−6	−14
Manitoba, Canada	*−4*	*0*	*15*	*38*	*52*	*62*	*66*	*64*	*54*	*41*	*21*	*6*
Montréal, Quebec	−11	−9	−4	5	13	18	21	19	15	8	1	−7
Canada	*13*	*15*	*25*	*41*	*55*	*65*	*69*	*67*	*59*	*47*	*33*	*19*
Helsinki,	−6	−7	−4	1	8	14	17	16	11	6	0	−4
Finland	*21*	*20*	*25*	*34*	*46*	*57*	*62*	*60*	*52*	*42*	*32*	*25*
Moscow,	−11	−9	−4	3	12	17	19	17	11	4	−2	−8
USSR	*12*	*15*	*24*	*38*	*53*	*62*	*66*	*63*	*52*	*40*	*28*	*17*
Barnaul, USSR	−18	−16	−10	1	11	17	20	17	11	4	−4	−12
	0	*3*	*14*	*34*	*52*	*63*	*68*	*62*	*51*	*35*	*17*	*6*
Kazan, USSR	−14	−12	−7	3	12	17	20	17	11	4	−4	−12
	7	*10*	*20*	*38*	*54*	*63*	*68*	*65*	*51*	*39*	*25*	*11*
Yakutsk, USSR	−43	−37	−23	−9	5	15	19	16	6	−9	−29	−41
	−46	*−35*	*−10*	*16*	*41*	*59*	*66*	*60*	*42*	*16*	*−21*	*−41*
Vladivostok, USSR	−15	−11	−3	4	9	14	19	21	16	9	−1	−10
	5	*12*	*26*	*39*	*48*	*57*	*66*	*69*	*61*	*49*	*30*	*14*
Harbin, China	−19	−15	−4	6	13	19	22	21	14	4	−6	−16
	−2	*5*	*24*	*42*	*56*	*66*	*72*	*69*	*58*	*40*	*21*	*3*

Differences in conversion are due to the figures quoted being originally in Fahrenheit.

Cold humid climate weather stations—Rainfall

Station	Average rainfall in mm (*inches in italics*)											
	J	F	M	A	M	J	J	A	S	O	N	D
Dawson	20	20	13	18	23	23	41	41	43	33	33	28
	0.8	*0.8*	*0.5*	*0.7*	*0.9*	*1.3*	*1.6*	*1.6*	*1.7*	*1.3*	*1.3*	*1.1*
Winnipeg	23	18	30	36	51	79	79	56	56	36	28	23
	0.9	*0.7*	*1.2*	*1.4*	*2*	*3.1*	*3.1*	*2.2*	*2.2*	*1.4*	*1.1*	*0.9*
Montréal	71	61	71	58	71	69	73	71	69	66	66	71
	3.7	*3.2*	*3.7*	*2.4*	*3.1*	*3.5*	*3.8*	*3.7*	*3.5*	*3.3*	*3.4*	*3.7*
Helsinki	46	36	36	36	46	46	56	74	64	66	64	61
	1.8	*1.4*	*1.4*	*1.4*	*1.8*	*1.8*	*2.2*	*2.9*	*2.5*	*2.6*	*2.5*	*2.4*
Moscow	28	23	30	38	48	51	71	74	56	36	41	38
	0.9	*0.7*	*0.8*	*0.7*	*1.2*	*1.8*	*2.4*	*2.4*	*2.2*	*1.6*	*1.2*	*0.9*
Barnaul	20	15	15	15	33	43	56	46	28	33	28	28
	0.8	*0.6*	*0.6*	*0.6*	*1.3*	*1.7*	*2.2*	*1.8*	*1.1*	*1.3*	*1.1*	*1.1*
Kazan	13	10	15	23	41	56	61	61	41	28	25	18
	0.5	*0.4*	*0.6*	*0.9*	*1.6*	*2.2*	*2.4*	*2.4*	*1.6*	*1.1*	*1*	*1.7*
Yakutsk	28	5	10	15	28	53	43	66	41	36	15	5
	0.9	*0.2*	*0.4*	*0.6*	*1.1*	*2.1*	*1.7*	*2.6*	*1.2*	*1.4*	*0.6*	*0.2*
Harbin	3	5	10	23	43	97	114	104	46	33	8	5
	0.1	*0.2*	*0.4*	*0.9*	*1.7*	*3.8*	*4.5*	*4.1*	*1.8*	*1.3*	*0.3*	*0.2*

Dry climates

The dry climates marked on the map, p. 81, fall into two broad types—semiarid and arid (or desert). The sole criterion for defining these climatic types is aridity, but aridity can be qualified by the effectiveness of the precipitation which is received. Parts of Western Australia have less than 250 mm (10 in) of rain a year but the low total falls at the right time of year for crops and it is also reliable—the area is therefore less arid than might be supposed and classed as subhumid. Conversely, there are places that receive more than 250 mm of rain a year but most of this falls in a few thundery showers and is lost by run-off on the hard earth. Such areas are therefore semiarid.

In general, a desert climate may be expected where rainfall is below 380 mm (15 in) a year, and almost always occurs where rainfall is below 250 mm (10 in) a year. Adjoining the deserts are the transitional semiarid areas, such as parts of the Sahel and Western Australia.

Deserts may be either hot deserts with no cold season (for example the Sahara in North Africa or the Namib desert in Southwest Africa) or cold deserts (for example the Gobi in East Asia) with a cold season—at least one month during which the temperature falls below 6°C (43°F).

Deserts and semiarid climates are caused by their position with regard to dominant high pressure systems, and thus most of the air masses reaching them are dry and there are few occasions suitable for rainfall. Some deserts are to be found right on the western coasts of continents—for example, the Namib Desert—and in these locations the prevailing winds are blowing from inland and therefore bring dry air and little to no rain. Patagonia in Argentina, a rain shadow desert (see p. 83), is a different case.

The Namib Desert on the coast of Namibia is one of the most arid places on Earth. Only the Atacama Desert in Chile is drier—the mean average rainfall near Calama is nil. (Telegraph Colour Library)

Dry (desert and semiarid) weather stations—Temperature

Station	Average temperature in ° C (*Fahrenheit in italics*)											
	J	F	M	A	M	J	J	A	S	O	N	D
San Diego	12	13	14	14	16	18	19	20	19	17	15	13
California, USA	*54*	*55*	*57*	*58*	*61*	*64*	*67*	*68*	*67*	*63*	*59*	*56*
Alice Springs,	29	28	25	20	16	12	11	14	19	23	27	28
Northern Territory	*84*	*82*	*77*	*68*	*60*	*54*	*52*	*58*	*66*	*74*	*80*	*82*
Australia												
Tehran, Iran	1	6	9	16	22	27	29	28	25	19	11	6
	34	*52*	*48*	*61*	*71*	*80*	*85*	*83*	*77*	*66*	*51*	*42*
Aden, Yemen	24	25	26	28	31	32	31	32	31	29	27	25
	76	*77*	*79*	*83*	*87*	*89*	*88*	*87*	*88*	*84*	*80*	*72*
Kashgar,	–6	1	8	16	21	25	27	24	21	13	4	–3
Xinjiang, China	*22*	*34*	*47*	*61*	*70*	*77*	*80*	*76*	*69*	*56*	*40*	*26*
Walvis Bay,	18	19	19	18	17	16	15	14	14	16	16	18
South Africa	*65*	*66*	*66*	*65*	*62*	*60*	*59*	*57*	*58*	*60*	*61*	*64*
Tarfaya,	16	16	17	18	18	19	20	20	21	20	18	17
Morocco	*61*	*61*	*63*	*64*	*65*	*67*	*68*	*68*	*69*	*68*	*65*	*62*

Differences in conversion are due to the figures quoted being originally in Fahrenheit.

Dry (desert and semiarid) weather stations—Rainfall

Station	Average rainfall in mm (*inches in italics*)											
	J	F	M	A	M	J	J	A	S	O	N	D
San Diego	46	48	38	15	8	3	3	3	3	10	23	46
	1.8	*1.9*	*1.5*	*0.6*	*0.3*	*0.1*	*0.1*	*0.1*	*0.1*	*0.4*	*0.9*	*1.8*
Alice Springs	46	43	33	23	15	15	10	10	10	18	23	33
	1.8	*1.7*	*1.3*	*0.9*	*0.6*	*0.6*	*0.4*	*0.4*	*0.4*	*0.7*	*0.9*	*1.3*
Tehran	41	25	48	36	13	3	5	0	3	8	25	33
	1.6	*1*	*1.9*	*1.4*	*0.5*	*0.1*	*0.2*	*0*	*0.1*	*0.3*	*1*	*1.3*
Aden	8	5	13	5	3	3	0	3	3	3	3	3
	0.3	*0.2*	*0.5*	*0.2*	*0.1*	*0.1*	*0*	*0.1*	*0.1*	*0.1*	*0.1*	*0.1*
Kashgar	8	0	5	5	20	10	8	18	8	0	0	5
	0.3	*0*	*0.2*	*0.2*	*0.8*	*0.4*	*0.3*	*0.7*	*0.3*	*0*	*0*	*0.2*
Walvis Bay*	0	0	0	0	0	0	0	0	0	0	0	0
Tarfaya	13	13	13	0	0	0	0	13	13	13	13	25
	0.5	*0.5*	*0.5*	*0*	*0*	*0*	*0*	*0.5*	*0.5*	*0.5*	*0.5*	*1*

* Walvis Bay receives virtually no rainfall per annum.

Mountain climates

Places at altitude experience cooler climates than those in the adjoining lowlands. The mountain climates marked on the map on p. 81 are the direct result of temperature loss with altitude. Local conditions may depend upon factors such as exposure to mountain and valley winds. Rainfall totals may be very low in valleys which are rain shadows but high on the intervening mountain ranges.

Examples of mountain climates include the Alps (which have a modification of the cool temperate climate), the foothills and lower ranges of the Andes (where modifications of the tropical climates of the adjoining lowlands are experienced) and the foothills of the Himalaya and the Plateau of Tibet (the former receiving a drastically modified version of the mild humid climate to the South and the latter with a version of the arid climate to the North). The High Andes and the upper parts of the Himalaya are so high that their climate almost resembles the polar type.

Mountain weather stations—Temperature

Station	Average temperature in °C (*Fahrenheit in italics*)											
	J	F	M	A	M	J	J	A	S	O	N	D
Davos in the Alps,	–7	–5	–3	2	5	10	12	11	8	3	–1	–6
Switzerland	*19*	*23*	*27*	*36*	*41*	*50*	*54*	*52*	*47*	*38*	*30*	*21*
Arequipa in the	14	14	14	14	14	14	14	14	14	14	14	14
Andes, Peru	*58*	*58*	*58*	*58*	*58*	*57*	*57*	*57*	*58*	*58*	*58*	*58*
Leh in the	–8	–7	–1	6	10	14	17	16	12	6	0	–6
Ladakh Range,	*17*	*19*	*31*	*43*	*50*	*58*	*63*	*61*	*54*	*43*	*32*	*22*
The Himalaya, India												

Differences in conversion are due to the figures quoted being originally in Fahrenheit.

Mountain weather stations—Rainfall

Station	Average rainfall in mm (*inches in italics*)											
	J	F	M	A	M	J	J	A	S	O	N	D
Davos	46	56	56	56	58	102	125	127	94	69	56	64
	1.8	*2.2*	*2.2*	*2.2*	*2.3*	*4*	*4.9*	*5*	*3.7*	*2.7*	*2.2*	*2.5*
Arequipa	30	43	15	5	0	0	0	0	0	0	3	10
	1.2	*1.7*	*0.6*	*0.2*	*0*	*0*	*0*	*0*	*0*	*0*	*0.1*	*0.4*
Leh	10	8	8	5	5	5	13	13	8	5	0	5
	0.4	*0.3*	*0.3*	*0.2*	*0.2*	*0.2*	*0.5*	*0.5*	*0.3*	*0.2*	*0*	*0.2*

This pattern of climatic belts is the result of the various factors outlined earlier in this chapter. These climatic zones are not a constant, and in Chapters 4 and 5 we saw that in past periods—and even in the relatively recent past—conditions were not the same. During the Little Ice Age, for example, the colder conditions seemed to have moved further to the South, in both the northern and southern hemispheres. The climatic belts have not always been exactly in the same position.

The gradual desertification of the Sahel (see Chapter 1) is one of the phenomena that is leading some scientists to suggest that another climatic change is happening. The freak weather conditions that have been experienced in Britain over the past decade—including the 'hurricane' of 1987 and the long hot summers discussed in Chapters 2 and 3—have also been interpreted by some as further 'evidence' of a climatic change. The reason for the change is said to be the greenhouse effect.

Meteosat image of the Earth, showing the water vapour content of the atmosphere on 15 June 1989. (European Space Agency/Science Photo Library)

7 The greenhouse effect

The Earth's atmosphere has many components. Oxygen and nitrogen are the principal ones but there are also other, much smaller, amounts of what have become known as 'greenhouse gases', the most important of which are carbon dioxide (CO_2), chlorofluorocarbons (CFCs), methane and water vapour.

Principal constituents of air	% by volume
Oxygen	21.0
Nitrogen	78.1
Water vapour	varies from 0 to 4
Argon	0.9
Carbon dioxide	averages about 0.3
Neon	under 0.002
Helium	0.0005
Methane	0.0002

Although the greenhouse gases may be present only in very small quantities, they have a vital part to play in maintaining our planet's 'heat balance'. These greenhouse gases act in a similar way as the panes of glass in a greenhouse. They let the Sun's heat through the atmosphere, but they do not let it all out again.

The heat from the Sun—solar radiation—controls our climate. It is not received equally over the Earth's surface, nor does all of the solar energy received in the exosphere (the highest layer of the atmosphere) penetrate to the ground. Some of the solar radiation is reflected off the 'edge' of the atmosphere back into space. The oxygen, ozone and water vapour in the atmosphere absorb some of the radiation, which warms the atmosphere—some of this energy radiates towards the Earth and is, in part, responsible for heating the surface of the planet.

Some solar radiation is reflected back into space off particles of dust, clouds and other matter in the atmosphere. The dense clouds over the tropical rain forests of Africa reflect solar energy, for example, while the relatively cloudless deserts and tropical grasslands to the North of them—in the Sahara and the Sahel—receive a far higher proportion of the radiation and thus reach greater temperatures during the day—although they do reflect a lot of radiation from the surface. (Conversely at night, when there are few clouds over the desert, there is a great fall in the temperature as energy is radiated back into the atmosphere.)

The radiation that reaches the Earth is not all used in heating its surface. Some is absorbed by the seas and oceans which act as a store of heat, giving it back again into the atmosphere in certain conditions. Very little of the radiation reaching a water surface is reflected—usu-

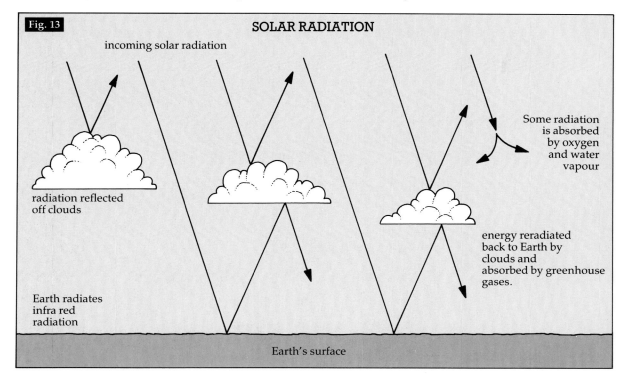

Fig. 13

SOLAR RADIATION

incoming solar radiation

Some radiation is absorbed by oxygen and water vapour

radiation reflected off clouds

energy reradiated back to Earth by clouds and absorbed by greenhouse gases.

Earth radiates infra red radiation

Earth's surface

ally less than five per cent—but when the Sun is low in the sky the amount of radiation reflected off water increases to up to 80 per cent. A varying, but considerable, proportion of radiation is either scattered or reflected from the Earth's surface. This is known as the albedo—literally the 'whiteness'. White clouds also reflect the Sun's rays back into space most efficiently. A dense bank of clouds can reflect up to 70 per cent of the energy received, while thinner clouds may reflect around 20 per cent. The white icecaps and snowfields are also good reflectors of radiation, and newly fallen snow can reflect up to 90 per cent of the solar energy received on its surface.

Different land surfaces produce different degrees of albedo. There is relatively little reflection off a landscape rich in vegetation. As little as eight per cent may be reflected off some forests (depending on the colour and shape of the leaves) or off some crops. On the other hand, the reflection of solar energy off a desert surface may be as high as 35 per cent.

Energy radiated from the Earth's surface returns to the atmosphere and some two thirds of this radiation is eventually lost to space. This reflection and reradiation is essential to the wellbeing of the planet which would otherwise become hotter and hotter.

The greenhouse theory proposes that the greenhouse gases—carbon dioxide, CFCs, methane and so on—are increasing in the atmosphere to such an extent that less reradiated energy is able to escape. The greater quantities of greenhouse gases absorb this long-wave radiation, effectively trapping it. The energy is then reradiated into the lower atmosphere. The gases may be said to be acting like a blanket, absorbing more and more radiation in the atmosphere. In theory, this could lead to a progressive increase in temperatures and eventually to climatic change.

NATURAL GREENHOUSE

Greenhouse gases seem to have been cast in the role of villains and in the alarm about the possible effects of the gases, according to the various theories of global warming, the *essential* role of these gases appears to have been forgotten. The greenhouse effect is not new or

The planet Venus which suffers a 'runaway greenhouse effect'. It experiences an average temperature of 464°C (867°F)—the heat received from the Sun is trapped beneath a thick cloud cover between 50 and 75 km (30 and 45 miles) above the surface of the planet. This view of Venus was taken from 720000 km (450000 miles) by Mariner 10's cameras on 6 February 1974. (Science Photo Library)

Over page The colour-coded topographic map is of the region surrounding Venus' north pole. The view is centred on the pole and extends down to a latitude of 20 degrees north. Yellow areas represent high ground and orange areas low ground. (USGS/NASA/Science Photo Library)

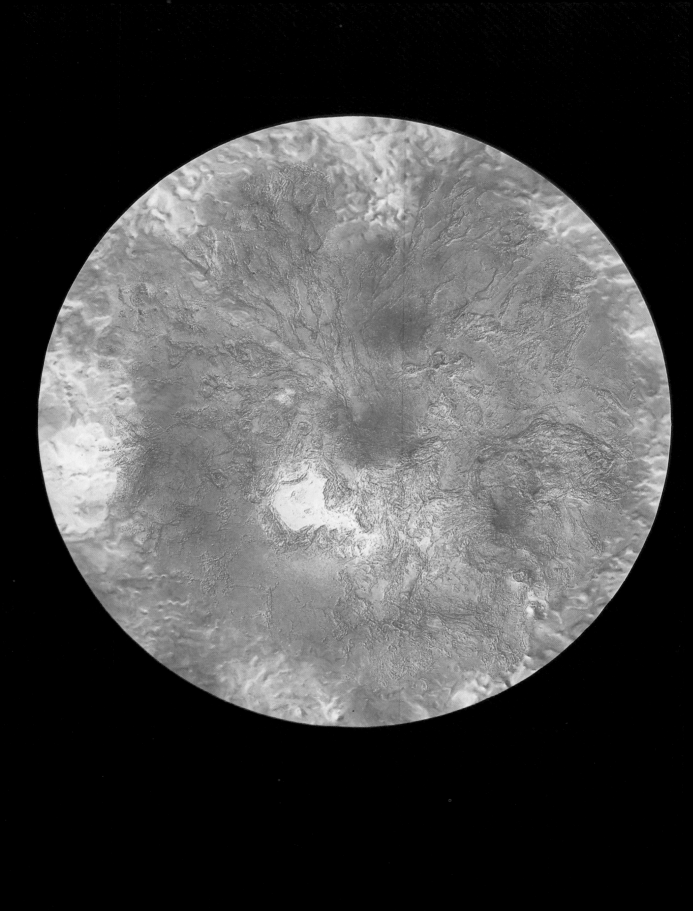

Above and right Carbon dioxide emission from factory chimneys and car exhausts. The amount of carbon dioxide in the atmosphere has increased by one quarter since the beginning of the Industrial Revolution. If unchecked, it will double within 100 years. (Spectrum Colour Library)

man-made. It is natural. Without a natural greenhouse effect there would be no life on Earth, whose temperatures would be far below the present global average of +14°C. If there were no greenhouse gases absorbing outgoing radiation, the loss of energy would be so great that the average temperature on Earth would be somewhere in the region of −15°C (5°F) according to some estimates or as low as −26°C (−15°F) according to others. There has been a natural greenhouse effect in the atmosphere of our planet for at least three billion years, and at some periods of geological time it is probable that it was far more powerful than it is today and may have been in line with—or even greater than—some of the most gloomy predictions of the greenhouse theorists.

GLOBAL WARMING

There is now general concensus that the world is warming. As recently as the 1970s such a proposition might have been disputed and there is still no agreement upon whether that warming is short-term or a more permanent feature. Although many sites are recording higher temperatures, the reasons for those increases may not be strictly climatic. Urban warming—the extra energy released during, and as result of, man's industrial activities, domestic heating, the use of the internal combustion engine and so on—has, for instance, had a measurable effect. Since the end of the 19th century some conurbations have warmed the atmosphere above them noticeably.

The general trend in global temperatures during the 20th century has been up. But such a bland statement obscures many underlying features, including the fact that during the 1940s and 1950s in the British Isles and in a number of northern hemisphere regions, including Greenland and the eastern seaboard of North America, average temperatures were declining. In the 1950s the Arctic icecap appeared to be advancing and in 1960s a theory of global *cooling* gained some academic respectability. It is, perhaps, as well to remind ourselves that we have come so rapidly from being concerned about global cooling, and the possibility of an imminent new Ice Age, to being alarmed about global *warming* and the greenhouse effect.

In the 1980s weather stations throughout the world recorded increases in temperatures. The arguments about *why* temperatures are rising are complex and intense. This rise is not an isolated phenomenon. Twice before this century, there have been substantial increases in temperature—between 1911 and 1917 and from 1926 to the early 1940s. Each of these periods was followed by a drop in temperature, although the overall trend was slightly up. Over the past century the overall increase has been about 0.5°C (0.9°F) which is considerably less than might have been expected from the rise of 30 per cent in the atmospheric content of carbon dioxide over the same period of time.

Could the previous temperature rises be ascribed to global warming? It would be difficult to make out a convincing case for the warm period of 1911–17, when the emission of greenhouse gases through the burning of fossil fuels was far lower, and had only just begun in earnest in much of the world as industrialization increased in the 1890s. It would have been far too early for any substantial carbon dioxide from such a source to have any climatic effect and the amounts involved would—by current standards—have been negligible. The tropical rain forests of Africa, Asia and South America, were largely intact (see p. 114), thus clearing and burning could not have been a source of more carbon dioxide in the atmosphere.

The existence of similar, though lesser, phenomena this century suggest that other factors might be at work, rather than just greenhouse warming. Nature is not simplistic and things are seldom as straightforward as they may at first appear. Nevertheless, some observers seem content to blame the effects of greenhouse gases alone for the current increases in warming.

MORE CO_2

Carbon dioxide (CO_2) is a colourless gas with very little odour. It has an acid taste. In structure it is a single atom of carbon joined by double bonding to two atoms of oxygen. Although carbon dioxide is most commonly found as a gas, it can also be found as a liquid and as a solid, the white 'dry ice'.

Carbon dioxide's natural cycle is tied to that of oxygen through photosynthesis in plant life (see p. 114).

Since the 1890s the amount of carbon dioxide in the atmosphere has risen by almost a quarter. At the turn of the century some 280 parts of carbon dioxide per million by volume were measured; the equivalent reading in unpolluted air in 1989 was in the region of 345 parts per million by volume.

This increase is overwhelmingly due to the burning of fossil fuels, especially oil and coal. The increased emission of carbon dioxide through the felling and burning of the tropical rain forests has not played nearly so great a role.

Other greenhouse gases are not present in such amounts as carbon dioxide but have increased at a faster rate. Methane, for example, is presently increasing at a rate of one per cent a year, which is not so alarming as it may sound as this gas is present in the atmosphere in only very small quantities—although this gas absorbs radiation more efficiently than CO_2. Chlorofluorocarbons (CFCs) are being emitted at an even greater rate. In the late 1980s CFCs in the atmosphere were increasing at an annual rate of six per cent. Again, this greenhouse gas is present in the atmosphere only in very small amounts but even small quantities of CFCs are thought likely to be capable of having a dramatic effect as they have been calculated to be 20 000 times more efficient than carbon dioxide in absorbing outgoing long-wave radiation.

Every year some five billion tonnes of carbon dioxide billow from the factory chimneys of industry and are pumped out of the exhaust pipes of cars and lorries. The burning of the tropical rain forests is thought to be adding as much as another billion tonnes of carbon dioxide per annum. The latter source of emission may be the easier to contain and steps are already being taken to curb the destruction of

Carbon dioxide in liquid form—dry ice. (Science Photo Library)

Over page The greenhouse effect. The build-up of certain gases in the Earth's atmosphere traps an increased amount of solar radiation and leads to a gradual warming of the whole planet, in much the same way that the panels of glass in a greenhouse trap the heat. (Science Photo Library)

Above Paddy fields are an important source of methane, a greenhouse gas. These paddy fields in Bali have the benefit of really rich soil, but overpopulation is severe and there is great pressure on the land. The size of the average land holding is only 1.2 ha (one acre). (Telegraph Colour Library)

Right Cattle are a major contributor to the methane gas in the atmosphere. The extent of cattle ranching has greatly increased since World War Two, particularly in tropical countries such as Brazil—the world's third largest beef producer—where vast areas of forest have been felled to create pastures. (Gamma)

the rain forest (see p. 112). On the other hand, our civilization needs energy and is heavily dependent upon fossil fuels. Although nuclear power is already being used as an alternative and research is continuing into the use of wave, solar and wind power, there seems little likelihood of reducing the world's consumption of fossil fuels in the near future. The Western world is, with varying degrees of commitment and success, making efforts to conserve energy, to become more energy-efficient and to seek alternatives, but the consumption of fossil fuels by Third World countries is gaining momentum.

METHANE

Methane gas too is increasing in the atmosphere as a result of the continued spread of agriculture to feed an ever-growing population. There are two main sources of the gas—paddy fields and cattle. The colourless and odourless methane is also known as marsh gas and is produced by the bacterial decomposition of vegetation under water that occurs in the flooded rice fields of South and Southeast Asia and the Far East. These fields are prepared for seeds and then flooded, allowing other vegetation to become established, even if it is only for a very short period. Weeds, too, may grow and eventually rot underwater. Rice leaves may die and rot submerged, while at the harvest, roots and stems may not all be removed. It is difficult to prevent at least some vegetation decomposing through bacterial action in the paddy fields, and as the crop feeds about one half of the world's population, and its cultivation is spreading into more and more marginal land under the pressure of population growth, the reasons for the extent of the increase of methane from this source can be appreciated.

Cattle, which are being kept in greater numbers than ever before, are another major source of methane. All ruminants, such as cows, have four 'stomachs'. Grazing cattle crop the grass as rapidly as they are able and, after mastication with their large tongue, pass it down into the rumen, which acts as storage chamber and then on to the reticulum where cellulose in the plants, which cannot otherwise be digested, is attacked by minute micro-organisms known as gut flora. The food is then regurgitated and the cud is chewed before being passed into the third 'stomach', the omasum—where water is absorbed—and finally into the true stomach, the abomasum. In this complicated process, the food is fermented, producing large quantities of methane gas as a by-product which the cow expels. The emission of methane gas by cattle is substantial enough to be considered a major contributory factor in greenhouse warming, and some 'green' observers have cited this as an additional reason for vegetarianism.

The large-scale farming and ranching that has developed over the past two centuries has greatly increased the number of cattle in the world. The extent of the problem of methane gas from cattle can be gauged from the fact that cattle outnumber people in many countries. Although other ruminants also add to the emission of methane gas, cattle—by virtue of their numbers and their particular digestive system—are the major contributors.

Not only is the amount of methane in the atmosphere increasing, but it is also not being broken down as rapidly as would normally be the case, owing to a greater concentration of carbon monoxide in the air.

CALCULATING THE EFFECTS

The Carbon Dioxide Information Analysis Center, at Oak Ridge, California, released figures in July 1989 which revealed that emissions of carbon dioxide—the principal greenhouse gas—rose substantially in 1987 to the rate of 5600 million tonnes per annum. This represented an increase of 1.6 per cent in a single year, and brought the rise in the period 1983–7 to 10 per cent.

Other sources disputed these figures, citing a steeper increase. The World Resources Institute, which is based in Washington DC, announced a 2.8 per cent increase in the emission of carbon dioxide in 1987 and a 3.6 per cent increase in 1988.

According to the Carbon Dioxide Information Analysis Center, the major contributors to the rise were China (with an increase of 4.8 per cent in 1987), the USA and the USSR. Britain, too, registered an increase of 2.6 per cent in 1987, although France and West Germany both succeeded in reducing their levels of carbon dioxide emission—France by 3.2 per cent (in part through a greater concentration upon nuclear power) and West Germany by 2.2 per cent.

It is not possible to predict future rates of emission of greenhouse gases, although models can be created to calculate the increase in heat for any given increase in the gases. This is, however, only part of the equation. The greenhouse theory predicts cumulative warming. For example, substantial increases in temperature on the Earth's surface would lead to greater evaporation from the seas and oceans, and therefore an increase in the amount of water vapour in the atmosphere. As the amount of vapour in the atmosphere varies considerably anyway, the results of a rise in this greenhouse gas have tended to be underestimated. However, water vapour is a particularly effective greenhouse gas and is one of the major contributors to global warming.

An increase in the amount of water vapour in the atmosphere would inevitably mean greater cloud-cover, as when water vapour rises it condenses to form dew, fog or clouds. Research indicates that cloud-cover worldwide has increased by 10 per cent during the 20th century, but it is debatable what climatic effect an increase in cloud cover might have. Some observers hold that an increased cloud-cover would exaggerate the greenhouse effect by trapping yet more outgoing radiation beneath it. Others forecast that more cloud will bring lower temperatures as incoming radiation will be reflected back into space off the clouds. There is probably no reason to suppose that either of these possibilities is mutually exclusive.

Other factors, too, have been put forward as components of a cumulative greenhouse effect. For example, large deposits of methane have been discovered on the seabed in recent years. They are believed to occur naturally as a result of the biological degradation of submarine vegetable matter such as seaweeds and other plants. This methane exists in 'bubbles' kept in place by the pressure of water above them. Some scientists have suggested that any warming of the oceans could disturb this fine balance and allow the methane to rise to the surface, thus adding considerably to the emission of one of the most effective greenhouse gases.

Also, when sea water increases in temperature it is possible that it will be unable to absorb so much carbon dioxide. At present the sea—and the life forms that live in it—absorb around 50 per cent of the carbon dioxide that is emitted into the atmosphere. The warmer water becomes, the less carbon dioxide it is physically able to absorb. Thus, it is argued, if global warming were to increase the temperature of the seas, the effect would become progressive.

TIME-LAG

We have seen that the global increase in temperature since the beginning of the 20th century is something in the region of 0.5°C, which is not nearly as great an increase in temperature as the rise in carbon dioxide over the same period of time would have suggested. This has led some observers to propose that the effects of any increase in greenhouses gases will not be as great as feared. But other scientists have forecast a time-lag between the emission of the gases and the consequent increases in temperature.

The major results of any substantial warming would, it is predicted, stem from an increase in the temperature of the oceans—the melting of the Arctic icecap, possible rises in sea level and a shift in the climatic belts. The oceans, though, would be unlikely to respond quickly to any such warming because of their immense scale. A time-lag of 20 to 50 years has been forecast before the full effects of global warming are experienced.

This time delay has been described as a climatic timebomb ticking away until the second quarter of the 21st century. Many scientists believe that the amount of carbon dioxide already released into the system, but which has not yet had time to have any repercussions, is enough to ensure a rise in temperature of at least 1°C by 2025, which may not sound much but seemingly small changes can have a considerable effect in the long term. As this estimate is based upon the assumption of no further emissions—a most improbable event because of man's economic dependence upon fossil fuels (coal and oil)—the actual rate of global warming predicted by most scientists is considerably more.

THE OZONE LAYER

Among environmental issues, the hole in the ozone layer is probably second only to the greenhouse effect in public consciousness. The ozone layer is a fragile belt of gas some 15 to 35 km (9 to 22 miles) above the Earth. It is composed of ozone gas—which is formed of three oxygen atoms, whereas oxygen is formed of two. Ozone gas is derived from atmospheric oxygen as a result of the effects of solar radiation. The ozone layer protects the flora and fauna on Earth by filtering out most of the ultra-violet part of the solar radiation received by the planet. Ultra-violet radiation can cause skin cancer, encourage the formation of eye cataracts, reduce humans' immunity to disease and severely damage crops. Without the shield of the ozone layer, the amount of ultra-violet radiation received on Earth would be multiplied almost one hundredfold. The result would be an ecological disaster on a scale almost too great to contemplate.

Between 1982 and 1985 two scientists working for the British Antarctic Survey—Joe Farman and Jonathan Shanklin—discovered a 'hole' in the ozone layer in the South Polar region. Their research confirmed that between 1955 and 1985 the amount of ozone in the atmosphere over the Antarctic had fallen by nearly one half.

In 1956, when the first measurements of the ozone level were made, there was no 'hole', but seasonal changes in ozone levels were suspected by George Dobson, who perfected a method for measuring the amount of ozone. (The units in which ozone is measured are now called Dobson units.) In 1956 Dobson took measurements at Halley Bay in Antarctica and found that in September and in October the ozone level was 150 Dobson units less than calculations suggested it should have been, but by November the readings were according to expectations. Dobson initially suspected faulty equipment, but the readings he took in the following year confirmed the pattern. Measurements

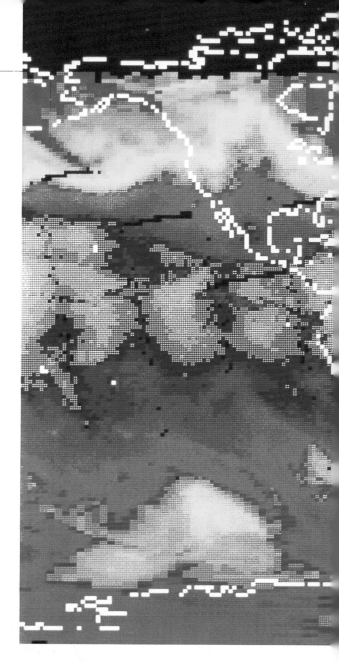

taken in other parts of the world did not repeat the pattern, but throughout the 1960s the seasonal decrease in ozone levels at Halley Bay were recorded. Then in the 1980s the hole was discovered.

Satellite photographs have revealed the development of a hole in the ozone layer over the Antarctic continent every spring since 1970. At first this phenomenon was largely ignored. The damage was contained to a relatively small area by the prevailing winds and the effects were minimal. But another hole was found over the Arctic and this, and the alarming rate

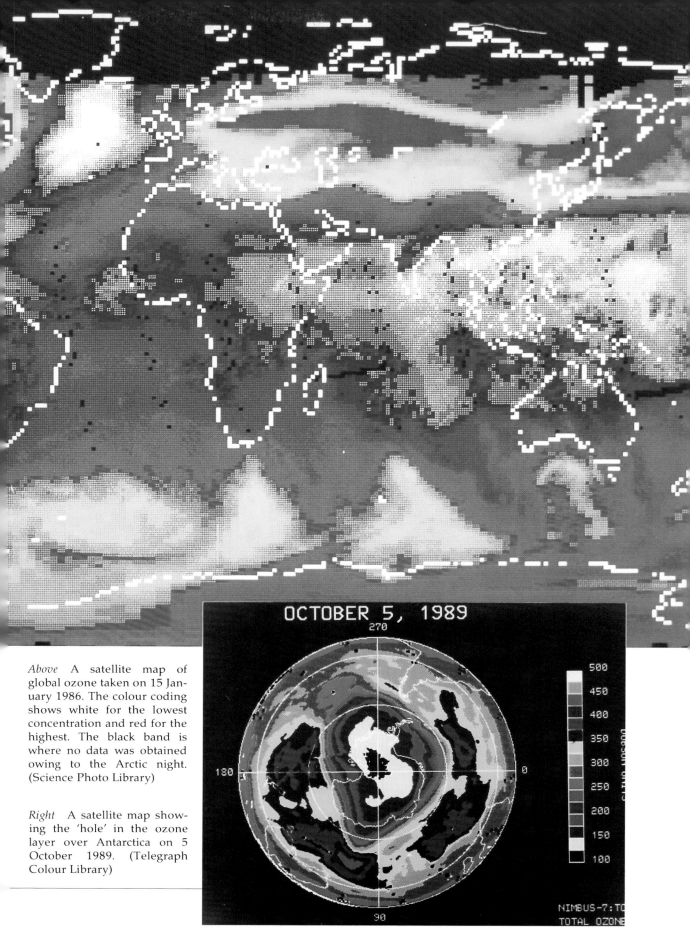

Above A satellite map of global ozone taken on 15 January 1986. The colour coding shows white for the lowest concentration and red for the highest. The black band is where no data was obtained owing to the Arctic night. (Science Photo Library)

Right A satellite map showing the 'hole' in the ozone layer over Antarctica on 5 October 1989. (Telegraph Colour Library)

OCTOBER 5, 1989

270

180

0

90

500
450
400
350
300
250
200
150
100

NIMBUS-7:TO

TOTAL OZONE

at which the Antarctic ozone level was falling, began to cause concern.

In 1990 scientists working for the Australian Bureau of Meteorology released figures which showed that ozone levels in the Antarctic had fallen by 10 per cent over the period 1987–89. Now there is growing evidence that there is a substantial reduction in ozone levels in the upper atmosphere—the stratosphere—throughout the world, although the decrease is not uniform. The decrease has been less in the equatorial regions where a reduction of about one per cent has been experienced since the 1950s. Throughout the whole of the rest of the southern continents—Africa, South America, Australia and Oceania—the decrease has also been slight, in the order of three per cent and at the most five per cent. A similar slight reduction in ozone levels has been experienced in the Caribbean, the southern states of the USA, North Africa and the Mediterranean, Arabia, India, China, Southeast Asia and Japan.

Outside these regions a more serious depletion has taken place. Across the northern part of the USA, Canada below the Arctic Circle, all of Northwest Europe and virtually the whole of the Soviet Union, the reduction in the ozone layer has been almost 10 per cent, while in both the Arctic and Antarctic the thinning of the layer has been in the order of 10 to 100 per cent.

The decrease is greatest towards the Poles and varies with the seasons. In most years the hole that appears over the Antarctic in spring is 'repaired' within a few months. These seasonal variations are caused by the fact that during the Antarctic winter there is no sunlight, and light is essential for the chlorine from CFC gases to break down the ozone. Thus for several months a year no such breakdown can happen and by spring a hole in the ozone layer has formed.

Data collected from satellites and ground weather stations show that the effect is spreading and that the Arctic decrease is almost enough to qualify as a hole. In 1989 the hole in the ozone layer over the Antarctic was bigger than ever and, for a time, enlarged and shifted to cover part of Patagonia and the Falkland Islands. And in 1989 and 1990 measurements taken in the Alps in Switzerland alarmed scientists with the first indications that thinning was progressing over Europe. The level of ultra-violet radiation received over the Swiss Alps has risen by about one per cent annually since 1980, according to the findings of a study made by the University of Innsbruck (Austria). These findings were, however, disputed by ozone experts at the United States National Institute of Health who had recorded a *decrease* in the amount of ultra-violet radiation received in some parts of the USA between 1974 and 1985. This apparent paradox has been explained by some as an increase in the amount of man-made ozone (and of other gases that absorb ultra-violet radiation) as a result of pollution of the atmosphere over major urban areas.

Nevertheless, the expectation is that holes in the ozone layer could appear at a number of locations in the northern latitudes, particularly in Europe and North America.

CFCs

The most likely reason for the global thinning of ozone is the increased emission of CFCs (chlorofluorocarbons), (see p. 103). CFCs are man-made gases that are used in a number of household gadgets and industrial processes including fridges, many aerosols and in some air conditioning. They are also used in insulation and some foam-blown cartons. The CFCs released from these items and in these processes cause damage in the upper atmosphere when they decay into chlorine gas which destroys ozone.

Much publicity has been given to 'ozone-friendly' products—some of which may reduce the emission of CFCs only fractionally—and upon efforts to help phase out CFCs through international aid to countries such as India and China and other less-developed countries in an attempt to establish substitute technologies in them.

The degree of public awareness of the issue was highlighted in a survey undertaken in 1990 in the United Kingdom by the Henley Centre which forecasts market trends. The survey showed that concern over the thinning of the ozone layer was greater—among a sample of the British public—than any worries about Third World poverty, cruelty to animals, water pollution or nuclear war. The report also showed that people in Britain were willing to pay extra for products that did not use CFCs.

Ozone layer and weather

The bulk of publicity concerning the ozone layer problem has focused upon the possible effects of an increase in the amount of ultra-violet received on Earth in terms of health (especially cancer) and in terms of diminished crops. But the thinning of the ozone layer could also have profound climatic effects.

Any increase in ultra-violet radiation is likely to reduce the amount of plankton in the oceans. The important role played by plankton with regard to the greenhouse effect is noted on pp. 156 and 157. Global thinning of the ozone layer could increase any such effects.

Additionally, changes brought about by thinning of the ozone layer could have an effect upon global wind patterns, with obvious repercussions upon world climate. The major concentration of ozone in the upper atmosphere is over the Equator—the region of greatest production of ozone gas. This zone is referred to as the ozone ridge. Any substantial thinning of ozone in the upper atmosphere could have an effect upon thermal patterns which could result in alterations in the wind patterns in the stratosphere. Thus, although the ozone problem is usually spoken of only in health and agricultural terms, it is interwoven with the greenhouse effect and world climatic change.

8 The rain forests

The tropical rain forests cover over 30 million sq. km (12 million sq. miles), that is over 20 per cent of the Earth's land surface. The tropical evergreen forest—that is the tropical rain forest proper—occurs within 4 degrees North and South of the Equator, although other varieties of tropical forest—for example tropical deciduous forest and tropical open woodland—are to be found covering wide areas between the Tropics and Capricorn and Cancer.

The vast sea of green leaves of the tropical forests plays a vital role in the climatic regime of the Earth. Chlorophylls in leaves drive the chemical reaction photosynthesis in which carbon dioxide enters the leaves through microscopic pores to be converted into sugar glucose. At the end of the reaction, oxygen is released by leaves, and thus photosynthesis is the source of all the oxygen in our atmosphere. In this way the tropical rain forests provide much of the oxygen upon which we rely for life itself, and the destruction of considerable areas of this forest for logging and farming is a great cause of concern.

In photosynthesis, quantities of carbon dioxide are used up. Thus, in a situation where plants and animals live together, the ratio is maintained in the existing levels of oxygen (which is taken in through respiration by living organisms and released through the leaves of plants) and carbon dioxide (which is released by living organisms but taken in by leaves). This fine balance is in serious danger of being upset by the felling and burning of large areas of tropical forest. The imbalance caused through the increased emission of carbon dioxide that would result from the destruction of much of the rain forest through clearing and burning would greatly increase the amount of greenhouse gases in the atmosphere and could produce global warming.

An immense amount of water is held in the ecosystem of the tropical forests. The tropical

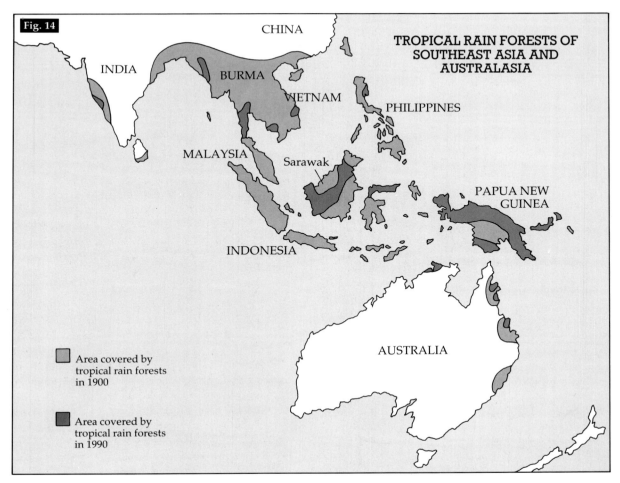

Fig. 14

TROPICAL RAIN FORESTS OF SOUTHEAST ASIA AND AUSTRALASIA

CHINA

INDIA

BURMA

VIETNAM

PHILIPPINES

MALAYSIA

Sarawak

PAPUA NEW GUINEA

INDONESIA

AUSTRALIA

Area covered by tropical rain forests in 1900

Area covered by tropical rain forests in 1990

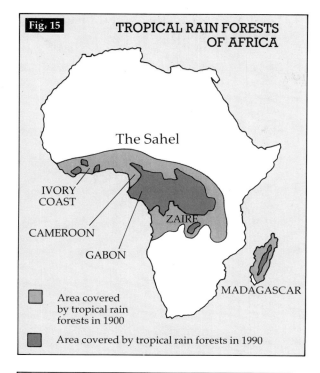

Fig. 15

TROPICAL RAIN FORESTS OF AFRICA

The Sahel

IVORY COAST

CAMEROON

GABON

ZAIRE

MADAGASCAR

☐ Area covered by tropical rain forests in 1900

■ Area covered by tropical rain forests in 1990

TROPICAL RAIN FORESTS OF THE AMERICAS

BELIZE

VENEZUELA

PANAMA

Amazon Basin

COLOMBIA

BRAZIL

☐ Areas covered by tropical rain forests in 1900

■ Areas covered by tropical rain forests in 1990

Fig. 16

rain forests receive at least 4000 mm (156 in) of rain a year, with no real dry season and little significant variation from one month to the next. The luxuriant growth of the vegetation of these forests forms a near continuous canopy of leaves and branches through which scant light can penetrate to the forest floor. Occasional so-called emergent trees break through above the canopy which is some 30 m (nearly 100 ft) high, and when a tree is felled or dies, there is fierce competition among the low young trees as they fight for the light to fill the gap of the fallen tree. Beneath the canopy the air is very humid. The plants of the tropical forest draw up water through their roots and transport it through the plant in the xylem (tissue). Eventually that water is evaporated or transpired through pores in the leaves. In the great heat of the tropics—with average temperatures often in excess of 27°C (80°F)—this humid air warms and rises, causing further precipitation.

The tropical rain forest is thus both a result and a determinant of climate. The destruction of the forest not only increases erosion and the rate of runoff but also greatly reduces the humidity and in turn the precipitation in the area.

THE RAIN FORESTS UNDER THREAT

The tropical rain forests are the 'lungs' of the Earth. Once the rich and varied forests of the tropics covered virtually all of the Amazon Basin in South America, the West African coasts, the basins of the River Zaire and its tributaries in central Africa, the islands of Indonesia and most of Malaysia and Papua New Guinea. Now, because of agricultural and other human pressure upon the land, vast swathes of equatorial forest have been cleared.

Clearing and burning is destroying a resource that cannot be replaced. The rain forests have an important climatic role to play. They are, at the same time, both the result of climate and a determining factor of climate. If the rain forests disappear a far-reaching climatic change will be inevitable. Between 1964 and 1984 some 13 100 000 sq. km (505 790 sq. miles) of forest and woodland were destroyed worldwide—that is an area larger than Canada and the State of Alaska combined.

The great tropical forests are an important part of the natural ecosystem of the planet. They store and release water, they supply oxygen to the atmosphere and, when burned, they are a major contributor of carbon dioxide to the atmosphere. Large areas of the Amazon rain forest are being cleared (above), for the construction of highways (left), for creating pasture to raise cattle (above right) and flooded when hydroelectric dams are constructed (right). (World Wide Fund for Nature and Gamma)

In some countries deforestation has been especially drastic. In the Ivory Coast, for example, 80 per cent of the rain forest was destroyed between 1960 and 1990. The accompanying maps show how much of the rain forest has been lost this century and the extent of the surviving forest.

The virgin forest is usually felled between April and June. The task is completed by fires during the following months. Land hunger is the driving force, but the fertility of the soils of the rain forest is illusory. The forest itself is thick and lush, but that is not a reflection of a particularly fertile soil. It is rather a perfect adaptation to the equatorial climate with its high temperatures and heavy rainfall.

Equatorial soils tend to be poor and lateritic. The forest cover affords them some protection, but, once cleared, the earth is exposed to severe erosion and leaching, which in lower gradient areas is the more important factor. The new farmland created by deforestation can support agriculture for only a couple of years. It is exhausted quickly and the topsoil is soon washed away by the heavy tropical rain. This is dramatically illustrated in Madagascar, where the greater part of the tropical rain forest has been cleared for farming. Gullying has occurred within two to three years as deep sharp-sided small valleys have been carved out and the soil has been carried away by streams and rivers to be redeposited in the Indian Ocean. The dense vegetation of the Madagascan forests retained much of the precipitation, releasing it at a steady rate to the fast-flowing streams of the island. The land was therefore well-watered all year. Now, the rivers flow unevenly as the run-off is rapid from the eroded countryside, and for part of the year the former rain forest suffers from drought.

BRAZILIAN RAIN FOREST

Brazil is gradually invading its great natural reserve, the Amazonian rain forest. In most years in the 1980s an area the size of Belgium was cleared and burned in the Amazon basin to create new pastures or farming land or to allow for the construction of hydro-electric power dams.

The majority—70 per cent—of the rural population of Brazil is landless. Opening the interior is, perhaps, the obvious 'safety valve' to defuse the Brazilian nation's land hunger. Every year many thousands of peasants flock up the highways of Transamazonia from the South, seeking land. New highways slash through the smouldering remains of what was forest and the title deeds to virgin land are easily obtained. Unfortunately for the peasants, the land is poor and their attempts to farm it do not allow them more than the very poorest living.

Gold fever has attracted thousands more to clear the forest and to threaten the very existence of the Amazonian Indian tribes. The protests of the Yanomami Indians, who had lived virtually untouched by the 20th century in isolated forested highlands, made world headlines. The Yanomami found their land invaded by gold panners and under threat from plans to construct dams which would have flooded important tribal areas. The Kayapo Indians, too, were able—with international support—to stop Brazilian plans to develop their part of the Amazon basin through the construction of massive dams.

During 1989 and 1990 attitudes towards the rain forest seem to have changed. Conservationists and environmentalists have campaigned vigorously on behalf of the Amazonian Indians and have actively promoted the fact that the rate at which the great tropical rain forest of Brazil is being destroyed has reached a level which is of global concern.

They have shown that the burning and clearing of the Amazonian rain forest has led to water pollution on a serious scale, that soil loss and erosion have increased drastically, that many species of plants and of birds and animals are threatened with extinction owing to the loss of their natural habitat, and that the emission of large amounts of carbon dioxide through the burning of the forest is adding to the greenhouse gases in the atmosphere at a rate that is causing extreme concern.

By the end of 1989 the efforts of conservationists were beginning to have an effect upon the continuing destruction of the rain forest in Brazil. The first serious curbs enacted by the government of Brazil against burning the Amazonian rain forest had resulted in a diminution of clearing. During 1989, 30 per cent less forest was felled in Brazil, a saving that was in part due to conservation legislation and in part due to a particularly long rainy season that prevented further slash and burn clearing.

IBAMA—the Institute of the Environment and Renewable Resources in Brazil—and the Brazilian Forestry Department have established helicopter patrols to check that conservation laws are being adhered to, but attempts to control the burning of the Amazonian rain forest have not been easy. Two enforcement helicopters have been attacked and a forestry official has been murdered by ranchers who felt that their livelihood was threatened by the measures that had been taken to protect the forest.

IBAMA has confiscated the chainsaws that had been distributed by one candidate for the governorship of Amazonas State, and has prosecuted a farmer in Rondonia State for setting fire to over 25 sq. km of forest. (Rondonia has an especially bad record in the destruction of the rain forest. By 1988 one sixth of the state or about 42 000 sq. km (over 16 000 sq. miles)—an area equivalent in size to Switzerland—had been felled in Rondonia.)

The Institute is also undertaking an education programme to encourage the conservation of the forest. However, Brazil's conservation laws are not yet strong enough to safeguard the survival of a precious and irreplaceable resource. Brazilian farmers are still legally empowered to clear up to half their land as long as the felling of trees does not endanger rivers or water sources. The laws are lax enough to have allowed over 250 sq. km (nearly 100 sq. miles) of forest to be lost in Acre State and over 200 sq. km in Rondonia in 1989.

ASIAN RAIN FORESTS

In Southeast Asia, too, there is much pressure upon the tropical rain forests. Some countries have taken measures to attempt to preserve their remaining forests. Thailand, for example, which has seen a very considerable reduction in extent of its tropical forests, has banned logging entirely, and both Indonesia and the Philippines have enacted legal restrictions upon the amount of timber that can be exported.

However, concern is especially great at the rate at which Malaysia is exporting tropical hardwood. The Malaysians are the world's leading exporters of tropical hardwood and to maintain this supremacy the country is permitting what many regard as the wholesale destruction of its forests. The Malaysian authorities are unreceptive to increasing international criticism on the subject, even when it comes from such quarters as the Prince of Wales in his Rainforest Lecture delivered at Kew in 1990.

In Malaysia the issue is seen in terms of trade and short-term economic policy rather than as a vital environmental concern. Some Malaysian observers are reported to consider the environmental lobby as a ploy by the Western developed nations—who include the most important producers of softwoods—to gain control of the world timber market. To others, the choice though is a stark one between vital revenue for a developing country and concern for the long-term effects upon climate and the environment of continued logging.

Malaysia's rain forests have been cleared at an astonishing rate. At the turn of the century, the area now covered by the Federation of Malaysia had nearly 320 000 sq. km (nearly 125 000 sq. miles) of forest. In 1950 the majority of the country was still forested. By 1989 only 80 000 sq. km (over 30 000 sq. miles) of virgin forest remained.

In peninsular Malaysia—the former Malaya—the Federal Government has taken effective measures to preserve what remains of the forest, and now there is very little felling. Perhaps some notice was taken of climatic data recorded in the hill stations of western Malaysia which has been interpreted by some as evidence of the greenhouse effect.

The felling of the tropical rain forests is regarded to be one of the major contributory factors of the greenhouse effect. If that effect is a reality, we should expect to find increases in temperature as evidence of global warming. Such figures have been recorded in the resorts of the Cameron Highlands in Pahang State in Malaysia.

The Cameron Highlands are a cooler plateau area, developed as a hill station by the British during the colonial era. Several degrees less than the surrounding equatorial lowlands, the Highlands rely upon their pleasant climate to support a resort trade and market gardening. The mean annual temperature in the Cameron

Below Amazonia in Brazil contains the world's largest concentration of tropical hardwoods. The Amazon Forest has been estimated to cover 330 million ha (81.5 million acres). (Gamma)

Left Fire claims a greater acreage of forestry every year than felling. (Gamma)

The Cameron Highlands in Western Malaysia are often cloudy and shrouded in mist. (Spectrum Colour Library)

Highlands used to be given as 18°C (64°F), but over the last 30 years the average temperatures in the cool hill stations have risen to 19°C (65°F). It is too soon to know whether this is a temporary climatic variation or a climatic change.

The Malaysian state of Sabah—in North Borneo—used to be swathed in a nearly continuous blanket of tropical rain forest. The attentions of land hungry farmers, mainly shifting cultivators clearing the forest by slash and burn, and of loggers, have decimated the stands of trees. By 1990 virtually nothing remained of the once endless sea of leaves and branches.

In the Sarawak, over 2800 sq. km (over 1000 sq. miles) of virgin forest—an area equivalent in size to the English county of Derbyshire—is being felled annually. So great is the economic motivation that gangs of loggers are working by searchlight through the night. Existing conservation legislation restricts the timber companies operating in Sarawak to felling less than half a dozen valuable species of tree. In practice the strict letter of the law is not always adhered to, and countless other trees are cleared or irreparably damaged in order to gain access to the commercial varieties.

If logging continues in Sarawak at the present rate, the state's stands of timber will no longer exist by 2020 and the timber industry, the mainstay of its economy, will have collapsed. Not only are the hill and swamp rain forests of Sarawak being decimated by logging, but also the fisheries of the state are under threat because of heavy pollution of its rivers caused by the activities of the timber companies.

The Sarawak forests are among the oldest stands of tropical trees in the world, and they support a rich variety of flora and fauna. In 1989–90 a group of conservationists and forestry experts, led by Lord Cranbrook, researched conditions in Sarawak and presented their findings in a report to the International Tropical Timber Organisation. They condemned the destructive scale of logging in the state and proposed a 30 per cent phased reduction in logging in order to conserve the rain forest. The report highlighted the possibility of the complete destruction of the forests of eastern Malaysia and the threat to wildlife and to local tribes.

The wildlife of the Sarawak rain forest includes endangered species such as the orang-utan, the flying fox, the hornbill, the proboscis monkey and the gliding snake. The disappearance of their habitat threatens them all with extinction. Only two per cent of Sarawak is currently protected, despite the declared official intention to set aside eight per cent of the state as reserves. The state government has the powers to introduce stricter controls for the environment but up to 1990 it has not used them.

If the activities of the timber companies and the lack of protection offered by the state government, are responsible for the disappearance of the rain forest, the traditional agricultural practices of local tribes such as the Kayan are not entirely blameless. These shifting cultivators clear the forest in order to grow their crops for a couple of years, until the poor soil becomes exhausted. Then they move to another stretch of virgin forest and by slashing the vegetation with their knives and burning the felled trees, they create new temporary clearings for farming. The advent of the chainsaw to Sarawak has increased the effectiveness of their clearance and intensifed the pressure that they are exerting on the land.

The Belaga district of Sarawak was, until 20 to 30 years ago, covered by a flawless dense well-watered rain forest, sustained by high temperatures and heavy reliable rainfall. Logging has reduced the area to bare, deeply eroded scrubland, punctuated only by occasional clumps of trees. The high humidity of the forest used to be drawn up by the heat and eventually fall again as rain, almost daily. The destruction of the tropical rain forest has interrupted this predictable weather pattern. In 1989–90 Belaga was hit by a severe drought for the first time. The Malaysian government had to rush food to the district to avert famine when the paddy fields dried out. It does not take too many steps to link the cause and the effect, the wholesale change to the environment and the drought.

It is pertinent to ask *why* the lowland forests, the mountainous jungles and dense vegetation of the swamps are being cleared in Sarawak. The timber is the mainstay of the local economy, but is it being used to good purpose? The bulk of timber exported from the state is imported into Japan where much of it is used not for furniture nor for construction but rather for the manufacture of cheap throwaway items including chopsticks.

BIOMASS BURNING

Burning the tropical rain forest to create new farming land is adding more than just carbon dioxide—the principal greenhouse gas—to the atmosphere. Research by scientists at the Max Planck Institute, at Mainz in Germany, suggests that the burning is contributing increased amounts of gases containing nitrogen. They believe that when a tree is consumed by fire, up to half of the nitrogen within it is emitted in molecular form.

In the nitrogen cycle, plants absorb nitrogen gas in a manner very similar to that in which they absorb carbon dioxide, and convert it chemically into a source of energy. The nitrogen is returned to the atmosphere through the action of bacteria in the soil as dead trees and other plants decay, thus completing the cycle. But when the trees are destroyed by fire, the return of the nitrogen is speeded up. Large amounts of nitrous oxide and nitric oxide are emitted.

Nitrous oxide is a much more effective greenhouse gas than carbon dioxide. It has been calculated to be 250 per cent more powerful than carbon dioxide as an agent capable of confining the Earth's heat within the atmosphere. (Nitric oxide is not a greenhouse gas but is nevertheless important environmentally. It is active in forming nitric acid, one of the main components of acid rain.)

This newly discovered link between the burning of the rain forest and the emission of nitrous oxide should be treated with some caution. Although the role of nitrous oxide as a greenhouse gas is known, the emission would occur naturally anyway through the actions of soil bacteria on decaying vegetation. It would, however, be substantially slower. We are not yet in a position to know whether the more rapid emission of gases that would be released as part of the natural nitrogen cycle may unbalance that cycle and increase warming, but some scientists feel that this is unlikely.

EXPLOITATION FOR CONSERVATION

The remaining stands of virgin tropical rain forest are unlikely to stay intact. Indeed, their very survival depends upon them being exploited *and* preserved.

Appeals to conserve the rain forest on grounds of aestheticism have little chance of success. The countries in which the remaining equatorial forests are found cannot be expected to preserve them merely because they are beautiful. Nor will the pleas of environmentalists on behalf of the orang-utan and the hornbill be sure to encourage loggers to stay their chainsaws.

The climatic case is more convincing, but the prevention of long-term climatic change—even if 'long-term' may be as little as 40 years—is not certain to influence those whose more immediate economic survival is at stake. It is simply not worthwhile for those who are living in or exploiting the rain forest to stop logging and cutting it down for subsistence farming in the short-term. The financial pressures upon them are too great to make it worthwhile to preserve the trees in order to prevent an uncertain climatic future halfway through next century.

If the tropical rain forest is to be saved, those who make their living from it will demand and expect compensation. As global warming is an international problem, so only international intervention is likely to be on a scale to tackle it.

International aid agencies and the governments of developed nations have realized that the wealthy West will have to pay the poor Third World in order to keep the rain forest. In 1990 the European Community summit in Dublin discussed the issue of the tropical forests and agreed to offer aid to the Brazilian government in an attempt to save the forest. The EC's plan envisaged debt relief packages

for Brazil's large crippling foreign loans in return for measures taken by the Brazilian government to preserve the remaining rain forest. These measures would probably include stricter forestry conservation legislation and codes of conduct for EC-based timber firms and timber exporting industries.

The government of Brazil indicated its willingness to consider the conversion of its foreign debt in return for conservation of parts of the Amazonian forest. President Collor's anxiety to reduce Brazil's 115 billion dollar foreign debts may have been the determining factor. The adherence of Brazil—the guardian of the largest rain forest—to the 'debt-for-nature' movement received a warm welcome.

Similar 'debt-for-nature' arrangements have already been made between the United States and a couple of Latin American governments, including those of Ecuador and Costa Rica. If the EC debt relief plan is successful it might be extended to European companies logging in Africa and Southeast Asia.

The conversion of debts and the ensuing legislation to ban logging may not by themselves be enough to ensure the future of the rain forests in Brazil or anywhere else. Most environmentalists believe that success will only be achieved when the people who live in and obtain a living from the forest are also

Below left and right The rapidly disappearing rain forests of Sarawak are not only an important natural resource but also the home to many threatened species, including the hornbill and the orang-utan. (Spectrum Colour Library)

given a financial incentive to conserve. The forests will only be safe when it becomes more profitable for local people to respect them than to cut and burn them down to create farmland.

It is, of course, possible to declare an area of rain forest to be a national park, protected from logging and clearing. But this cannot be the solution for all the threatened forests. The problem is too big. Nor does it take into account the practicality of protecting the rain forest from the land-hungry people and from small-time loggers. The areas involved are too big to police, regardless of the huge expense of attempting to do so.

It is probably only feasible to conserve some of the rain forest in national parks. The remaining extent of the forest, around the parks, could be 'managed' for limited logging for high value timber and used for 'agro-forestry' in which only the smaller trees are felled. In the shady environment thus created, plants such as coffee could be cultivated commercially. Such a system of management would preserve the high canopy of trees which would prevent soil erosion and any climatic change—such as has happened at Belaga in Sarawak (see p. 123)—in the area.

Proper management of the forest will require a great deal of aid and training which could come from companies as well as aid agencies and governments. Managed forest could include 'bio-diversity' parks maintained by international chemical companies. The plants of the rain forest form a bewildering reserve of chemical compounds that can be extracted for drugs for many medical purposes. For example, a compound in one tree in West Africa is believed to kill certain cancer cells.

MORE TREES

The felling of the tropical rain forest is removing one of nature's most efficient ways of reducing the amount of carbon dioxide in the atmosphere. Through photosynthesis, plants absorb this greenhouse gas, and by cutting and burning trees the emission of carbon dioxide is increased.

Large-scale reafforestation—both tropical and temperate—has been proposed as a bio-logical way of stalling global warming. Seven million square kilometres (over 2 700 000 sq. miles) of new forest, it has been estimated, would be sufficient to absorb all the carbon dioxide currently emitted by the burning of fossil fuels.

Some schemes have already begun to encourage tree planting, although the extent of forest destroyed each year is greater than the area of new plantations. In temperate regions, the forests are being damaged not only through clearance but also by the effects of acid rain, another result of the pollution of the atmosphere by industry.

PLANTING

Restoration of forests is also being attempted. In Britain, for example, there are several major schemes for afforestation, including 'urban forests' such as the proposed discontinuous forest to be planted in suburban Southwest Essex. New stands of conifers are being planted, with some controversy, in ecologically important moorland and marshy areas in Caithness and Sutherland in northern Scotland and in 1990 plans were approved for the plantation of a large new forest in Leicestershire and Staffordshire in the English Midlands.

In Guatemala, a new forest is being created, directly linked to an industrial project. Applied Energy Services of Virginia, USA—better knopwn as AES—is paying for the planting of 52 million trees. It has been calculated that a new forest of this size would be able to absorb the equivalent amount of carbon dioxide to the emission from AES's new power station in the USA.

Reafforestation has an important part to play in helping to combat global warming. It is possible that major forestry projects worldwide could diminish the effects of global warming through absorbing the increased quantities of carbon dioxide emitted into the atmosphere. If temperatures increase enough, extensive new forests could be encouraged in what are now the frozen wastes of the tundra. These regions are vast enough to be able to support huge new forests capable of absorbing enough carbon dioxide to halt or slow down global warming, though their soils are poorly developed and unsuitable for forests at present.

ACID RAIN

The burning of fossil fuels, and certain industrial smelting processes, causes the emission of sulphur dioxide, a pungent acidic gas. In 1989 it was estimated that some 150 million tonnes of sulphur dioxide were released into the atmosphere in this manner, a much greater total than the relatively small amount of the gas emitted naturally during volcanic eruptions. Atmospheric sulphur dioxide is gradually oxidised and and when combined with atmospheric water vapour, this becomes acid rain—in this case dilute sulphuric acid. Coal-fired power stations, and a number of industrial processes, also release nitrogen oxides into the air and these, too, when combined with moisture in the atmosphere, become another component of acid rain, dilute nitric acid.

The problem can be seen at its most severe in the temperate forests, especially in parts of Switzerland, Sweden, Britain, Czechoslovakia, Norway and Germany. Large stands of trees in the Black Forest have been destroyed by acid rain. The Germans refer to the problem as *Waldsterben*, literally 'tree death'; in few places are its effects more evident than in northern Bohemia (Czechoslovakia) and in the adjoining areas of Thuringia and Saxony in Germany. Here the technological base of industry under the former Communist regimes of Czechoslovakia and the German Democratic Republic (what was East Germany) was not as mindful of the environment as in some Western countries. The air was heavily polluted by old-fashioned factories and the emission of sulphur dioxide and nitrogen oxide was large. Today, driving from Prague to Leipzig, a motorist will pass the blackened, dying skeletal remains of pine forests on the crests of the hills, a concentrated and graphic example of the effects of acid rain. Britain, though, has a higher percentage of trees damaged by acid rain—up to 67 per cent—although the results are seldom as far advanced as they are in much of Germany and Czechoslovakia.

The long-term effect of the diminution of the temperate forests caused by acid rain could be a sizeable increase in the amount of atmospheric carbon dioxide. Preventive measures are being taken though, as filter systems to stop the emission of the acid rain gases are being fitted to power stations and industrial plants.

Trees damaged by acid rain in Slovenia. (Gamma)

9 High tide

If the Earth were to warm owing to the greenhouse effect, one of the most dramatic results that has been forecast is the melting of the vast polar icecaps. Research in both the Arctic and the Antarctic in the period 1960–1990 suggested that this process may have already begun.

ANTARCTIC ICE

Observations by the British Antarctic Survey base at Rothera in the South Orkney Islands reveal a substantial loss of ice in the area, the result of a sizeable increase in temperatures recorded at the site over the last quarter of a century. The mean temperature recorded at Rothera between 1982 and 1986 was 1.1°C (34°F) which compares with only 0.1°C (32°F) between 1977 and 1981.

There is little doubt that the edges of the ice caps in Antarctica are retreating, in some places more rapidly than at others. The retreat recorded in the South Orkneys is especially dramatic and correlates with the marked increase in summer temperatures. But is it the result of global warming?

Scientists working at Rothera have evidence that warming and the consequent retreat of the ice has been going on since about 1950, which would suggest that if the greenhouse effect is the culprit, the phenomenon has been active on a measureable scale for at least 40 years. Alternatively, it may mean that other factors, such as cycles of solar activity (see p. 170), have been at work. Research by the British Antarctic Survey has shown that there have been similar retreats, associated with similar warming, in past periods when there was a smaller icecap and considerably more vegetation in the South Orkneys than there is now.

There is gathering evidence of cycles in climate. The scientists at Rothera suggest that a natural cycle of temperature is now at a peak. By the turn of the 21st century, it is suggested, the natural trend in temperature in Antarctica should be downwards. Yet by the year 2000 the greenhouse effect is likely to be evident, cancelling out the opposite natural trend. There is great concern that, over the next few years, greenhouse warming could be building up on top of natural warming. Global warming is expected to be more intense in the polar regions than in other parts of the world (see p. 140), thus an acceleration of the retreat of the ice is expected.

Left A radio operator at the British Antarctic Survey base of Rothera. Daily radio schedules from Rothera are maintained with Cambridge via a satellite link. (Science Photo Library)

Right Icebergs on the edge of the Polar icecap. (Images)

Previous page The Thames Barrier—a tidal barrier constructed between 1973 and 1983—was designed to reduce the danger of flooding in London. Any substantial rise in sea level as a result of global warming would test the barrier. (Spectrum)

Over page Vabbinfaru on Kaafu Atoll in the Maldives, one of the low-lying tropical island groups thought to be at risk from inundation in a greenhouse world. (Spectrum)

ARCTIC THINNING

The Scott Polar Institute in Cambridge announced in 1989 evidence of thinning of the Arctic icecap. This was seized upon immediately by some observers to be firm proof of global warming owing to the greenhouse effect.

British nuclear submarines took sonar readings under the Arctic icecap between 1976 and 1987 that indicate a thinning of the icecap, in an area immediately North of Greenland, from an average of 6.7 m (22 ft) at the beginning of the survey period to 4.5 m (14.8 ft) in 1987. The change was a dramatic one, far greater than any previously recorded—another fact that was emphasised by proponents of greenhouse warming. It should be remembered, though, that accurate measurement of the thickness of polar ice has been possible for a very short period of time indeed—far too short a time to formulate theories concerning average rates of melting or accretion. The thickness of polar ice observed in the past 30 to 40 years has varied from year to year, but the decrease since 1976 has been dramatic. That melting is not, though, by itself, firm proof of the greenhouse effect.

Concern about the melting of the polar icecaps has been growing, with fears that significant melting could lead to a rise in sea level and the inundation of low-lying areas. As noted elsewhere, the rate of melting, and the consequent rise in sea level, are hotly disputed by scientists with some proposing as little as 0.2 m (0.7 ft) by 2050 and others forecasting as much as 1.5 m (4.9 ft) by that date.

Both polar icecaps are melting. Some scientists contend that the Antarctic is more vulnerable than the Arctic, claiming that it is markedly thinner. They calculated that the Antarctic icecap is, on average, less than a metre deep. Most of the Antarctic ice is supported on land above sea level, thus any significant melting would be likely to result in a rise in the level of the oceans. The Arctic icecap averages a depth of 4 to 5 m (13–16 ft)—much of it the frozen surface of the Arctic Ocean.

In some places the Antarctic icecap is very thick indeed and for this reason it is thought, by some scientists, to be very stable and in far less danger of melting. The Arctic icecap floats on water and is thus held to be more at risk. Both icecaps contain large quantities of carbon dioxide, the emission of which would increase with any dramatic melting. This would, it is forecast, speed up the greenhouse effect.

PARADISE DROWNED

Global warming has so far been minimal, but scientists claim that computer models indicate greater increases of temperature in the 21st century, although the rate and the amount of that increase is hotly disputed. Any significant warming would be enough to begin to melt the polar icecaps. There is evidence that the polar ice in the Arctic and the Antarctic is thinning, but it is too early to attribute this to the greenhouse effect.

It has been calculated that if the emission of greenhouse gases continues at the present rate, the increase in temperature will be enough to cause a rise in sea level, owing to the ice caps melting, of between 24 and 38 cm (9–15 in) by 2030. A sea rise of about a third of a metre may not sound very much but to low-lying areas it could be devastating.

Few places are as threatened by such a rise in sea level as the 1190 coral islands that make up the Republic of the Maldives in the Indian Ocean. This remote archipelago—a former British protectorate—has had an uneventful history. Except for a single coup attempt that was foiled by Indian intervention, the Maldives have had a very quiet past. Their future, however, looks uncertain and as the islands celebrated the 25th anniversary of their independence in 1990, some writers could not help wondering just how much of the archipelago would be left above sea level in 25 years time.

The highest point in the Maldives is only 3 m (nearly 10 ft) above the ocean and that estimate of a rise in sea level of a third of a metre by 2030 is regarded by many scientists as conservative. A rise of less than a metre could submerge much of the island chain beneath the waves.

In November 1989 President Gayoom of the Maldives hosted a conference of leaders of small states that would be gravely endangered by any rise in sea level. The conference set up an action group composed of Trinidad and Tobago, Malta, Mauritius and the Maldives to inform governments of the crisis they perceive. Of the four nations in that group, the Maldives would face the greatest hardship from *any* rise in sea level.

The 200 000 Maldivians live on 203 of their 1190 beautiful coral islands. In less than a century it is possible that the entire Maldivian nation may be refugees if their low-lying island republic is inundated by the rising waters of the Indian Ocean. Its beaches which attract nearly a quarter of a million tourists annually, its coral reefs, its coconut palms may have all disappeared. One large wave has already flooded the capital, Male.

Computer models of possible effects of global warming indicate a profound change in the climate in the Indian Ocean. The Maldives currently experience a pleasant tropical climate with heavy rain brought by the monsoon between May and August. The tropical storms that strike much of the Indian Ocean largely bypass the Maldives.

The future could be violent and more unstable. Storms of increasing ferocity are predicted if global warming increases and these islands could become subject to hurricanes during which high waves could pass right over the top of the flat Maldives. A combination of the erosive force of the waves during repeated devastating storms and a rise in sea level could mean that the Maldives could eventually become the memory of an exiled people.

The Maldives—and other low-lying island groups of the Indian Ocean—are not the only candidates to become modern versions of the legend of Atlantis. Several Pacific Ocean island nations are equally at risk, foremost among them Kiribati and Tuvalu. Formerly known as the Gilbert and Ellice Islands, these two tiny independent Commonwealth nations are among the least known countries in the world. Kiribati is composed of three small groups of coral atolls set in over five million sq. km (two million sq. miles) of ocean. Although the single island of Banaba (formerly Ocean Island) attains a height of 81 m (265 ft), the remaining islands do not rise above 8 m (25 ft).

Tuvalu consists of nine small islands whose combined area is only 26 sq. km (10 sq. miles), and whose highest point above sea level is 6 m (20 ft). Its 8300 inhabitants make a poor living, raising pigs and poultry and gaining copra from the coconut trees that cover these archetypal South Sea islands. The people of Tuvalu have learned of the fate that many scientists predict for them if increased temperatures next century raise the sea level and, apparently, they just do not believe them.

Tonga, Micronesia, the Cook Islands and many other Pacific Ocean islands are thought to be in similar danger.

FLOODING

Some forecasters have predicted a rise in sea level of several metres before the end of the 21st century. Most of the scientists who concur with the global warming theory would dismiss such a scenario as an overestimate. Even if temperatures were to rise by more than 0.5°C (the increase experienced during the 20th century), it is thought unlikely that a partial melting of the polar icecaps would result in more than a slight rise in sea level—although this could be enough to threaten the low-lying island nations of Kiribati, Tuvalu and the Maldives. If a rise in the oceans of around a third of a metre (*c.* 12 in)—and this is the most common estimate—were to occur by 2030, many low-lying areas of the world would be vulnerable. Much of the Eastern seaboard of the United States could face inundation by the waves. Some coastal districts of Texas, Louisiana and Florida are at risk from a rise in the water level, and their problems are compounded by the natural subsidence that they are presently experiencing. Some parts of the Gulf of Mexico are subsiding by as much as one centimetre (nearly

half an inch) a year as the North American continent continues to adjust after the melting of the ice sheets at the end of the last Ice Age, which is now thought by some scientists to have been about 18 000 years ago.

A fear of flooding is an inbred part of the Dutch national psyche. The saying that 'God made the world, but the Dutch made Holland' is often quoted, and most of the provinces of North and South Holland, the most populous regions of the Netherlands, were made by man, reclaimed from the sea or from coastal brackish marshes. These areas were always marginal and Pliny wrote that in 'this eternal struggle in the course of nature it was doubtful whether the ground belonged to the land or the sea'. The Netherlands are protected by a series of sea walls and great dams that are continuously monitored and improved. The Delta Plan in the Southwest, the dams in the Rhine distributaries and the Afsluitdijk that holds back the Wadden Zee to protect the freshwater Lake IJssel and the entire new province of Flevoland (reclaimed from the sea since 1950) have changed the map of the country. All this would be at risk from global warming. It is,

The Delta Plan in the Netherlands involved the construction of a network of dams and dykes to protect the Southwest of the country from flooding. Begun in 1958, the scheme has sealed all the sea entrances except the western Scheldt and the approach channel to the port of Rotterdam. (Gamma)

Right Flooding at Khartoum, Sudan. Third world countries would be particularly threatened by the risk of flooding in a 'greenhouse world'. (Gamma)

Left and below Floods in Bangladesh. The threat to low-lying heavily-populated Bangladesh from the rise in sea level expected in a greenhouse world has been well-publicized, but the danger of flooding from inland—by the Rivers Ganges and Brahmaputra and their distributaries —is as great. It is thought that the run-off from the Himalaya will be much increased as climatic belts are shifted to the North, thus Bangladesh faces a double threat of inundation. (Gamma)

therefore, not surprising that the Dutch have been in the forefront of research into global warming and any consequent rise in sea level. Plans have already been drawn up to raise the height of the dams and sea walls to protect the 27 per cent of the country that lies below the present sea level and which houses almost two thirds of the nation's inhabitants. Dutch engineers express confidence that they would be able to beat off the challenge of a rise in sea level of as much as two metres (6.6 ft), that is an increase considerably above that forecast by the majority of greenhouse experts.

Other, poorer nations may be less fortunate, and it is Third World nations that seem to be especially at risk from flooding. Intense storms and high winds bring serious flooding to Bangladesh almost every year and in 1988 the homes of some 25 million people were destroyed or damaged and their fields washed away. The greater part of Bangladesh is covered by alluvial plains in the deltas of the Rivers Ganges and Brahmaputra, which combine as the Padma. Millions of Bangladeshis live in the swampy plains that are generally less than nine metres (30 ft) above sea level. The lowlands of Bangladesh are vulnerable to the possible effects of global warming in several ways. A rise in sea level is the first, and most obvious, threat. But a very substantial

increase in the flow of the rivers entering the country, and in the precipitation falling upon it, are also predicted. If sea levels do rise, there would be increased extents of ocean from which evaporation could take place and, perhaps, more rainfall. And the probability of flooding is also heightened by the very process that is thought to be contributing to global warming—deforestation.

The widescale clearance of forests in the Himalaya throughout the drainage basins of the Ganges and the Brahmaputra is leaving further areas exposed to erosion. Valuable topsoil is being washed away, landslides are becoming frequent and runoff more severe, so that downstream, in Bangladesh, the extra flow in the rivers is resulting in more flooding, although this could end up in the delta to add to the land area of Bangladesh.

Flooding on a large scale—as a result of heavier than normal rainfall—is growing throughout the world. There has been a threefold increase in floods of this type since the beginning of the 1960s, with a concentration in the developing countries of the Third World.

At a first glance this might be interpreted as the result of greater rainfall, but there is a gathering suspicion that these floods have been encouraged by deforestation in the developing world. The case has not been proved but the theory runs as follows. The very heavy rainfall of the tropics does not normally run off immediately. The natural vegetation of the rain forests breaks the force of the downpours and the plants, especially the trees, retain much of the water which is then released back into the water cycle only quite slowly. If loggers and farmers destroy the forests, erosion gathers pace and the precipitation drains away swiftly, causing floods downstream. In the longer term, this felling is thought to be capable of changing the local climate too, for example in Borneo (see p. 123).

FLOATING ICE

The outlook for Bangladesh, the Netherlands, Tuvalu, Kiribati, the Maldives and even New York and London would appear to be bleak with regard to flooding if the greenhouse theorists are right. Yet according to some theories, global warming may not lead to *any* rise in the level of the oceans.

The Arctic icecap is vulnerable, floating as it is on water. The rate at which most observers claim it is melting is rapid and if that suggested rate were to be maintained, the ice around the North Pole would have disappeared within a century. It is believed by some scientists that the Antarctic icecap is more stable and less likely to melt (see p. 130).

The Antarctic icecap is upon land and, if it were to melt, would cause a rise in sea level. The Arctic icecap floats. If it alone were to melt it should have little to no effect upon sea level at all. (It is, perhaps, useful to think of the analogy of a glass of iced water. When all the ice has melted, the level of the water in the glass remains the same.) If the theory forecasting the melting of only the northern icecap were to be correct, then no flooding should occur but the climatic effects would be startling.

SUNKEN SHORES

Most forecasts, however, combine a rise in temperature with a rise in sea level and those areas most at risk from flooding by the rising waves have already been identified. They include the Fens and large coastal areas of East Anglia, Lincolnshire and Humberside, stretches of low-lying land beside the Thames Estuary, Romney Marsh, much of the coastal plain of West Sussex and Hampshire, the low-lying areas of central Somerset, and most of the plains of Lancashire, Cheshire, the Solway Firth lowlands and the Carse of Gowrie.

If the rise in sea levels were to be about 30 cm (12 in) in the next century—a prediction that has become routine on the part of greenhouse theorists—this would represent a rise in water levels between three and six times above that with which existing British sea defences could cope. But this is only the initial rate of increase according to many scientists who predict that the sea levels around the coast of Great Britain will rise by between 10 and 15 cm (4 and 6 in) each decade if the greenhouse effect takes off.

The single most vulnerable area of Britain would be the Fens of Cambridgeshire and Norfolk. In July 1989, at a conference on the greenhouse effect organized by the Centre for Agricultural Strategy, the then British Minister of Agriculture, Fisheries and Food (Mr John MacGregor) observed about the Fens that 'unless we get things right, they may no longer exist'. Ironically, much of the Minister's own Norfolk constituency has been listed as an area in danger from rising sea levels.

A network of monitors measuring the tides around Britain has already been set up and this will keep a special watching brief on any rise in sea levels. If such a rise is detected, a costly system of coastal defences would be necessary to keep the sea out of low-lying areas. In 1988 some £28 million were spent on British coastal defences; by 1992 this figure is scheduled to rise to nearly £35 million. The Institute of Terrestrial Ecology has estimated that Britain would have to spend an extra £5 billion to build the additional sea defences to prevent the sort of flooding that would accompany global warming.

LONDON FLOODS

If sea levels do rise substantially as a result of the melting of polar icecaps many great cities would be at risk from flooding, including Venice, Rotterdam, Amsterdam, Bombay, Alexandria, Copenhagen, New York, Leningrad and London.

Fig. 17

'GREENHOUSE' BRITAIN

The dark areas are those most likely to be submerged by a rise in sea level of several metres.

Carse of Gowrie

Forth Estuary

Clyde Estuary

Solway Firth Lowlands

Tyne Coastal Lowland

Tees Estuary

Wigtown Bay

Furness

Morecombe Bay

The Fylde

Humber Estuary

Mersey Estuary

Dee Estuary

Lincolnshire Coast

The Fens

Norfolk Coast

Suffolk Coast

Carmarthenshire Coast

Somerset Levels

Essex Coast

Thames Estuary

Poole Harbour

Romney Marsh

Chesil Bank Lowlands

West Sussex Coast Plain

Solent Estuaries

The extent of the problem to be expected with global warming, according to one estimate in a University of East Anglia report, could be as much as 38 cm (15 in) by 2030. This would mean that the Thames Barrier, built to protect London from flooding, could be required to cope with high tides 20 cm (8 in) above the 1991 levels. At the same time the adjustment of the British landmass following the Ice Age is causing the Southeast of England to sink, thus the actual level of increase in high water level could be greater.

If a rise in sea level of this extent were to happen, 194 sq. km (75 sq. miles) of Greater London could be in serious danger of flooding by the middle of the 21st century. The critical point could be reached as soon as 2005, but that does not mean that flooding is inevitable from that point. However, the probability of the Thames Barrier being breached by floods would increase and the barrier would have to be reconstructed to minimize the effect.

A GREENHOUSE WORLD

If the icecaps melt, the resultant climatic changes would be likely to have an effect upon us all.

If the icecaps melt . . . Most recent models of the possible climatic changes resulting from the greenhouse effect suggest that only the Arctic icecap would be likely to melt. The Antarctic icecap, which in places is much thicker than the Arctic icecap, is held by some to be a more permanent fixture.

Hermann Flohn, of the International Institute for Applied Systems Analysis (the IIASA) has postulated an Earth with lop-sided climatic bands. He envisages a new meteorological Equator, some distance to the North of the true Equator, and an ice-free Arctic. The Antarctic, he forecasts, would continue to be covered by ice.

If one assumes a growing greenhouse effect of global warming—a scenario which is not yet

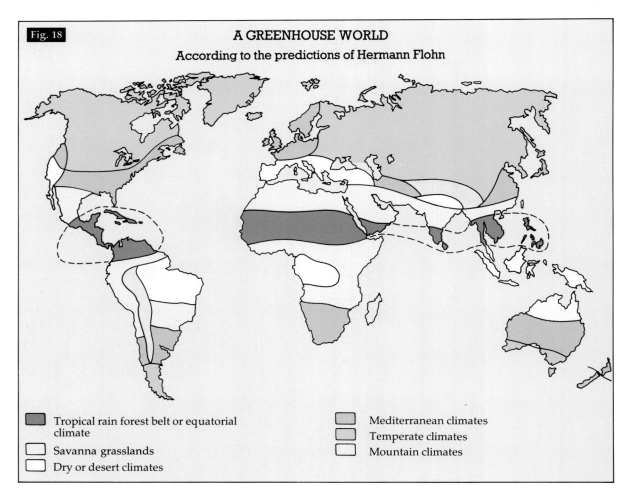

Fig. 18

A GREENHOUSE WORLD

According to the predictions of Hermann Flohn

- Tropical rain forest belt or equatorial climate
- Savanna grasslands
- Dry or desert climates
- Mediterranean climates
- Temperate climates
- Mountain climates

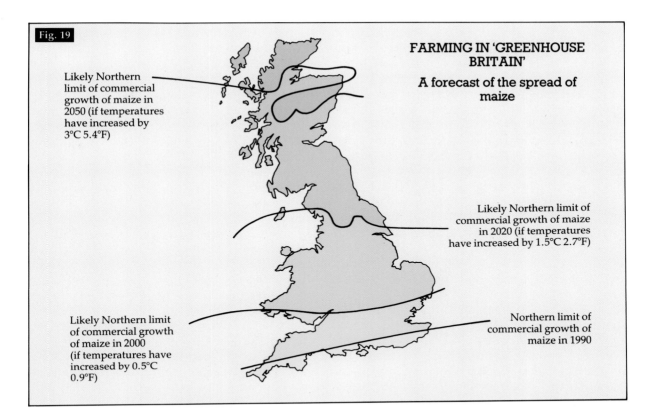

Fig. 19

FARMING IN 'GREENHOUSE BRITAIN'

A forecast of the spread of maize

Likely Northern limit of commercial growth of maize in 2050 (if temperatures have increased by 3°C 5.4°F)

Likely Northern limit of commercial growth of maize in 2020 (if temperatures have increased by 1.5°C 2.7°F)

Likely Northern limit of commercial growth of maize in 2000 (if temperatures have increased by 0.5°C 0.9°F)

Northern limit of commercial growth of maize in 1990

proven—the Arctic icecap could have melted by 2090.

As we have seen, the icecap in Greenland has melted, retreated and then advanced again dramatically in historic times. We do not know if the current period of dramatic melting is a rerun of climatic patterns from the recent past or the beginning of a new man-made phenomenon which is being imposed on top of a natural cycle.

Hermann Flohn appears to believe that the key to the present—and the future—does lie in the past. He has studied the climates of past periods to forecast the possible effects of global warming. He noted that the Arctic icecap is relatively new, in geological terms, having been formed when the northern continents assumed their present configuration around the North Pole some two and a half million years ago. The greater icecap of the Antarctic is over ten million years older.

It would take only an increase in temperature of some 4°C to melt the newer Arctic icecap. If this were to happen, the biggest immediate effect would be upon the temperature of the air masses above it. The current average winter ground temperature in the Arc-

tic is in the region of between −30°C and −35°C (−22°F and −31°F). The ice of the Arctic in part reflects the Sun's energy, but the open sea that would replace that ice would absorb the solar heat. It has been calculated that the average temperature of an open Arctic Ocean would be about +2°C to +5°C (36°F to 41°F), that is an increase in temperature in the region of almost 40°C. The effects of such a major increase would be dramatic and the entire weather pattern of the northern hemisphere would be changed.

According to Hermann Flohn, air temperatures would increase on average by around 11°C during the winter and 3°C in the summer. This would result in a shifting to the North of all the climatic belts in the northern hemisphere to resemble the pattern some three million years ago when the Arctic Ocean was ice-free.

The equatorial climate would shift North substantially to include Central America, the West Indies, the Sahel of West Africa, much of Sudan, Yemen, central and southern districts of India, Sri Lanka, Southeast Asia and the Philippines. This zone would receive the high temperatures and heavy rainfall, typical of the present equatorial zone (which is now slightly North of the Equator rather than on it because there are more land masses in the northern hemisphere than in the southern hemisphere). This would suggest a very different landscape in such parched areas as the Sahel, although it is unlikely that any new tropical rain forest could establish itself, especially with the probable pressure on the land that would result from population movements.

To the North a new savanna grassland might cover parts of Mexico, the Sahara, Saudi Arabia and North India. California, Texas, the Mediterranean countries, most of the Middle East, Turkey, Iran, the Punjab and the Ganges Valley would experience a substantial reduction in their winter rainfall and could become deserts.

In the southern hemisphere, too, the climatic belts would shift to the North, and while much of the present belt of tropical rain forest could become tropical grassland, there would also be a wide new belt of deserts which could include parts of the Amazon Basin and Northeast Brazil, most of the Zaire Basin, Malaysia, Indonesia and Papua New Guinea, plus northern Australia.

If such a shift of climatic belts were to occur—and Spain, Portugal, southern France, Italy and the Balkans were to become desert or semi-desert—the 'Mediterranean' climate would be pushed North, away from the Mediterranean Sea and would cover northern Europe, including Britain. This is one, probably extreme, scenario. It assumes that the current warming is the result of greenhouse gases, rather than the result of other factors (such as a climatic cycle dependent upon solar activity). It also assumes the melting of only one polar icecap.

'MEDITERRANEAN' BRITAIN

If much of Britain were to experience a Mediterranean type of climate, as most greenhouse theories forecast, there would be radical changes both in the British climate itself and in the agriculture that would be possible in the country. In a 'Mediterranean' Britain, the South could expect a January average temperature of about 6°C to 7°C (43°F) and a July or August average of nearly 25°C (77°F) or higher, with daytime summer temperatures regularly rising to over 35°C (95°F). The rainfall maxima would be in late autumn and during the first half of winter with just over 100 mm (4 in) in each of the months of November, December and January. The months of midsummer would be likely to be a time of drought with under 25 mm (1 in) of rain in some months.

The northern counties of England could experience a climate like that presently enjoyed by western central France and, in time, Scotland could be similarly blessed. The Atmospheric Research Group, based at the University of Birmingham, envisages Mediterranean crops being grown in southern Britain during the first half of the 21st century. Farmers would be obliged to respond to climatic changes and by 2030 maize and sunflowers could be common crops in most of England (see p. 46). The Group also predicts that the pastoral lowlands of western Britain will have to adapt to a sizeable diminution of rainfall and will have to adopt the type of arable farming currently associated with East Anglia. Northern Britain could become considerably wetter and warmer, and the dairy industry could be forced by climatic changes into the North and into those parts of Scotland, where soils are suitable.

Some people have imagined that vineyards may in future occur as far North as Central Scotland but the changes predicted are not all so attractive. If the upland areas of Wales and Scotland came into cultivation, with the climatic changes, there would probably be a dramatic increase in the erosion of the poor soils of those areas. Lowland Britain, too, could become subject to serious erosion as the country no longer experienced the typical fast fine rain and drizzle that so characterises British weather and increased thunderstorms, bringing sudden violent downpours of rain, could become more common.

A change in climate, and an emphasis on different types of crop, would also have an effect upon the varieties of weeds and pests in the country. It is probable that many of the benefits that might be derived from having a longer, warmer and drier growing season in the South of Britain and a longer, warmer and wetter one in the North, would be negated by the likely increase in pests, weeds and diseases.

MOSQUITOES

Mosquito nets were part of every Empire-building Victorian explorer's kit. We tend to think of mosquitoes as belonging to the tropics, but those hill top medieval towns that gaze down on the Italian coast were built on high to avoid the marshes where the malaria-carrying mosquito lurked. We do not expect to find mosquitoes in Britain, and when one recently hitched a lift on an aircraft to Gatwick Airport and then bit a Sussex publican it made newspaper headlines. But the return of the malarial mosquito to England has been seriously forecast as a result of global warming.

If the temperatures in the South of England increase by relatively little, it is possible that the mosquito could establish itself in areas such as the Fens, Romney Marsh and the Essex coastal marshes. If there had been any malarial mosquitoes in southern England during the hot August of 1990, the temperature was almost high enough for them to have been able to breed. Only the night temperatures dipped below the critical point. The common type of malaria carried by mosquitoes needs a constant temperature of 16°C (61°F) for between 10 and 14 days. The more deadly variety needs constant temperatures in excess of 20°C (68°F) for a similar period. Medical teams already make regular inspections of marsh areas in southern England that could with not too large a climatic change become the malarial mosquito's breeding grounds.

The mosquito (*Aedes*). The freshwater marshes of Romney Marsh and Somerset Levels could conceivably become breeding grounds for the mosquito in a greenhouse world. (Science Photo Library)

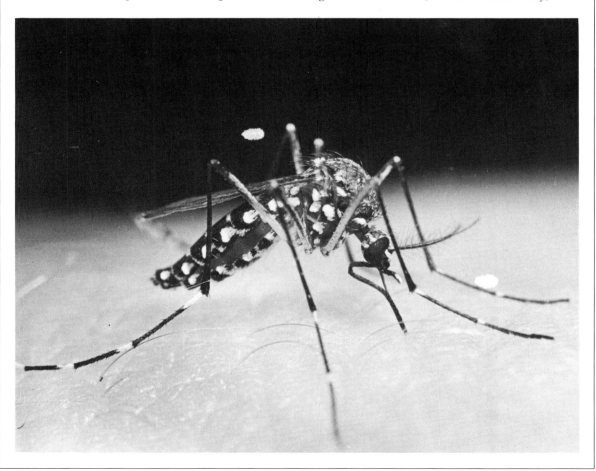

10 Cold comfort

Are we faced with an inevitable climatic disaster? If we were to believe the many, often alarmist, reports that have appeared in the popular press we would be drawn to assume an imminent global catastrophe. We know that the emission of carbon dioxide, methane and other greenhouse gases is increasing despite mounting concern and initial curbs taken by some Western governments. It is, therefore, only human nature to seek comfort from theories that minimize the possible repercussions of the greenhouse effect.

It has, for example, been suggested that the Earth's atmosphere and oceans are capable of absorbing any global warming that may result from the increased emission of greenhouse gases. The theory runs like this. Any overall rise in temperature will mean an increase in the warming of moist air masses over the oceans. These warm damp air masses will—if buoyant or forced to rise—ascend to a level at which the moisture-laden air forms clouds. Global warming will thus lead to an increased cloud cover which will, in turn, help to nullify the greenhouse effect as a greater proportion of the Sun's energy will be reflected back into space. By means of this self-corrrecting mechanism—the albedo (see p. 99)—the temperature will then fall again. The theory is attractive in its simplicity, but is thought by many scientists to be unlikely to work.

SECOND OPINION

Some may look for comfort to the recently revised opinions of Mikhail Budyko, the Soviet scientist who claims credit for warning the world about the greenhouse effect some 20 years ago. Mikhail Budyko is a member of the United Nations committee, set up to advise world leaders on global warming; but, despite his eminence and the attention given to his earlier theories on the greenhouse effect, relatively little publicity was given to the view he expressed in August 1989 that the greenhouse effect could be largely beneficial. He painted an appealing picture of deserts turned into productive pasturelands and of semi-arid Soviet Central Asia, supporting a flourishing arable agriculture as a result of increased rainfall. Budyko predicted that only a moderate increase in temperatures and rainfall would occur in the foreseeable future—increases far less than those envisaged by most observers.

Mikhail Budyko forecast a better world of bigger harvests and a consequent alleviation of famine. He went as far as to pronounce 'Paradise can return' and dismissed gloomy predictions of lowlands being submerged by the rising sea level and vast tracts of land on the edges of the tropics becoming desiccated. Budyko revised his own previous warnings of serious climatic change after studying fossil records to reconstruct transformations of the climate in past geological periods. However, the Soviet scientist's computer models—forecasting substantial increases in rainfall in desert areas and little change in temperate latitudes—are at odds with the models of virtually all his colleagues. Mikhail Budyko's views cannot, however, be dismissed. They will only be proved or disproved by events.

MORE SNOW?

What is not in doubt is that the higher temperatures resulting from the greenhouse effect will mean increased precipitation in many parts of the world. The warmer air becomes, the greater is its capacity to retain moisture. Thus global warming will increase cloud cover which will lead to greater rainfall totals in many places and substantially increased snowfall in the polar regions. Some forecasts predict that snowfall will be so much heavier that the polar icecaps will increase in size rather than melt with warming. This theory holds that the polar icecaps will endure and that low-lying

Previous page Tundra landscape in spring. Tundra regions have a brief summer during which the ground is free from snow, allowing a little vegetation to grow. But beneath the topsoil are frozen layers that never melt. (Spectrum Colour Library)

Facing page In the 1970s there was concern about the

possibility of another 'Little Ice Age'. Predictions of long cold winters and frozen estuaries and ports were common, although no one imagined anything quite so picturesque as this scene on the Hudson River in New York in the 19th century. The frozen river was a source for ice for domestic use and was 'ploughed' to cut blocks for storage.

areas are in no danger of inundation from rising sea levels. However, almost all the computer models—and the findings of scientists such as those attached to the British Antarctic Survey (see p. 130)—envisage that the biggest increases in temperature will be in the highest latitudes. The increases would be especially great in the polar regions, because, in the models, the seasonal snow is greatly reduced which would have a very strong effect on temperatures.

Some scientists are scaling down their estimates of the extent of global warming that might occur in the short-term. It is possible that the level of temperature increase will be slight for as much as 50 years, and perhaps even for as long as a century. But most scientists believe that the increase is real and in time—unless measures are taken to reduce the emission of greenhouse gases—global warming may have increased to a degree that a 'runaway' greenhouse effect may happen. The Earth simply cannot absorb heat indefinitely without the consequences of long-term climatic change.

RUNAWAY GREENHOUSE

If the possible consequences of the greenhouse effect were not worrying enough, some scientists have proposed the scenario of a 'runaway greenhouse'. They believe that, as global warming occurs, the increase in temperature will have a serious effect upon that vast treeless zone that covers much of the Northern Hemisphere between the frozen wastes inside the Arctic Circle and the coniferous forests of Canada and the USSR. This treeless zone with its permanently frozen subsoil is known as the tundra and as temperatures increase its character could change.

There has been concern that—with global warming—the tundra might 'dry out', releasing into the atmosphere great quantities of methane which is a strong greenhouse gas (see p. 106). Methane gas is capable of trapping heat in the lower atmosphere more effectively than other gases and is thus a serious potential contributor to global warming.

Methane is converted from carbon present

in the soil. It has been proved that about 15 per cent of all soil carbon is held in the frozen soil of the tundra, and at present methane released from the tundra accounts for some 10 per cent of the methane gas in the Earth's atmosphere.

Methane is present in the atmosphere at a ratio of 1.7 to one million, but the concentration of the gas is being increased by one per cent per annum, largely as a result of agricultural activities—from paddy fields and cattle (see p. 106). Climatologists have suggested that the higher temperatures associated with global warming will greatly increase the amount of methane gas released from the tundra and that this effect will be cumulative. Until 1990 it was assumed that this 'runaway greenhouse' effect would occur. However, the studies of W. S. Reeburgh and S. C. Whalen of the University of Alaska suggest that the reverse may be true.

Through analysis of soil samples in the tundra of the Aleutian Islands off the coast of Alaska, Reeburgh and Whalen showed that in drier tundra soils the release of methane gas decreased rather than increased. In experiments in which methane was injected into chambers containing moist—rather than waterlogged—tundra soil, it was found that nearly 55 per cent of the gas was absorbed by microbes in the soil.

The meadow plants that currently flourish in waterlogged tundra areas are efficient transporters of methane gas into the atmosphere. In these conditions the soil bacteria do not have the opportunity to oxidise the methane. But in less wet conditions the oxidising bacteria has greater scope to work.

If the tundra were to dry out with global warming, the water table would fall. This would have the effect of reducing the depth of waterlogged soil near to the surface which is the home of the bacteria that produces the methane gas. The drier soil created by the fall of the water table would result in conditions in which oxidising bacteria are present and active.

This oxidisation of methane in the soil would reduce the amount of the gas that would be released into the atmosphere. In what it is tempting to regard as a natural correcting mechanism, it seems that the increase in temperature from global warming would result in a reduction of the emission of methane gas

from the tundra, rather than the worrying 'runaway greenhouse' effect that has been suggested.

The one disadvantage is that if the methane is oxidised in the drier, but still moist, tundra conditions this would result in an increase in the emission of carbon dioxide, another greenhouse gas, albeit one which is only five per cent as effective as greenhouse agent as methane.

There is, unfortunately, an additional complication. To the North of the tundra stretch the great frozen wastes of permanent snow and ice caps. This 'snow desert' is to be found in Alaska, Canada, Greenland and the USSR. Methane lies trapped beneath its surface too.

If global warming increases, this expanse of permanently frozen land would warm too. The snows and ice would melt. And what was a snowy waste could become a waterlogged tundra, with conditions similar to those that were previously experienced to the South. It is a possibility that a new, wet tundra could come into being and that this would be an additional source of methane gas emissions into the atmosphere. We can only guess that the problem would remain the same, except that the source of the greenhouse gas would have moved northwards. At best, the replacement of one waterlogged tundra with another would be a very slow process and the emission of methane from this source is likely to either decrease or remain at a constant.

GLOBAL WARMING OR URBAN WARMING?

The increase in temperature of $0.5°C$ $(0.9°F)$ so far this century is real—but is it the result of the greenhouse effect? Figures from around the world confirm that temperatures are increasing, but not everyone is convinced about the cause of that warming.

As we know, climate is not a constant. It has changed and varied throughout the history of planet Earth, and those variations seem to come in identifiable cycles and patterns, that have been correlated with sunspot activity (see p. 170). The relationship between solar activity and the weather experienced on Earth has been demonstrated to the satisfaction of many climatologists and other scientists. It may be that observers have assumed that recent phenomena such as temperature increase, the melting of the polar icecaps and so on are the result solely of the greenhouse effect and have ignored the possibility that the conditions being experienced are evidence of a natural climatic variation, similar to those observed in the past. What could be a normal cycle may have been interpreted as a phenomenon that cannot be arrested.

We have seen that satellite readings of temperature have not yet recorded any significant increase in temperatures. This has been interpreted by some to suggest that there is no warming. Another interpretation has been that what has happened is *urban* warming rather than global warming.

The majority of climatic recordings have been taken near cities and other built-up areas. Temperature readings from major cities are invariably higher than those from physically comparable sites in the surrounding countryside. The industrial activities of man raise the temperature appreciably and it could be that observations taken from urban areas, or sites near to urban areas, have lead scientists to exaggerate any warming, although in the United States this has been carefully checked and only data not affected by urban areas have been used.

There are even arguments about the reasons for any increase in the amount of carbon dioxide in the atmosphere. We have become used to reading that the principal reasons for any increase in carbon dioxide are emission during the burning of the rain forests and the burning of fossil fuels (coal, oil etc). The situation is not, however, quite as simple as that and some scientists have gone as far as to propose that 'greenhouse' theorists are mixing up cause and effect. The counter argument holds that, as warmer air generally contains more carbon dioxide, higher temperatures are the principal *cause* of the increase in carbon dioxide in the atmosphere rather than being the *result* of increased carbon dioxide. A few observers use this theory to deny the existence of the greenhouse effect *in toto*. The truth may lie somewhere between.

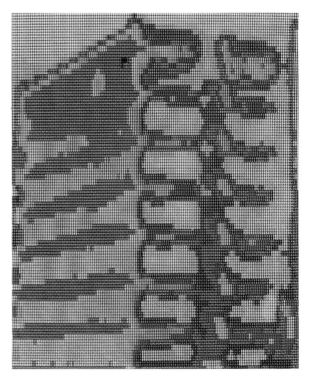

Urban areas are warmer than surrounding areas because of the heat given off by man's economic activities. On the left a photograph of an office block can be compared with a thermogram of the same block on the right, showing the distribution of temperatures across the exterior of the building. The Sun has heated the right side of the building, which is colour-coded yellow and red. The front of the building is not affected by the Sun and appears in green and purple. (Science Photo Library)

WHAT DO THE FIGURES SAY?

In October 1989 the United Nations Environment Programme Environmental Data Report revealed increases in global temperatures and in the gases that are associated with the greenhouse effect. The report showed that global temperatures have increased by between 0.3°C and 0.7°C since 1900. Global warming has not, however, been a constant. In some years this century global temperatures have fallen spectacularly—for example, during World War One, several years during the 1950s and as late as the 1970s. But overall the trend is up. Global temperatures are still increasing, and the report identifies 1981, 1983 and 1987 as especially warm years.

The authors of the report were not able, however, to put the blame for global warming on increased concentrations of greenhouse gases, but they did recognise that the warming has not been 'out of step' with the increase in these trace gases so far.

Over the 40-year-period 1947–86, data gathered by the authors of the report demonstrates that global warming has not been even. South America, parts of Africa, Australasia and the Antarctic have experienced the most significant increases, and warming has been greater over land areas than over the oceans. Some areas in the northern hemisphere have yet to show any warming and a few even registered a net cooling over the period of the survey.

The report also highlights a marked difference in the changing rainfall patterns of places in the higher and lower latitudes. High latitude regions—that is those between 35°N and 70°N, a band which includes Britain, northern and central Europe—have in general experienced increases in rainfall over the period of the survey. In the lower latitudes—that is in India and Southeast Asia and in the Sahel region of West Africa—the amount of rain received has decreased.

According to the survey, the levels of greenhouse gases have gone up dramatically. The report states: 'It is generally accepted that increases in atmospheric concentrations of greenhouse gases will cause significant increases in global temperatures and other climatic changes within the next century.'

Increases in both carbon dioxide and methane, the two principal greenhouse gases, were noted. Scientists measured the amount of carbon dioxide trapped in polar ice and concluded that levels had soared. The amount of methane trapped in the ice has doubled in the past 350 years, with the greater part of that increase coming in the last two centuries.

CLIMATIC RESEARCH UNIT

In 1990 the Climatic Research Unit at the University of East Anglia declared that the seven warmest years globally, since accurate meteorological records were first kept at the end of the 19th century, were 1980, 1981, 1983, 1986, 1987, 1988 and 1989. The Climatic Research Unit, in declaring the 1980s as the warmest decade on record, was drawing on data from weather stations around the world and from weather ships at sea.

The 18-strong unit has been producing computer models of weather systems and studying data on the possibilities of global warming for some years. Then a 'hurricane' in Southeast England in October 1987, prolonged drought in the summers in 1989 and 1990 over much of Britain and a growing awareness that the 1980s had been remarkably warm, heightened public interest in the greenhouse effect. The work of the team at the University of East Anglia became news. Politicians and world leaders quoted their findings and their theories and the reputation of the team, led by Professor Tom Wigley, was soon international.

Since 1979 they have been studying the greenhouse effect and in the eyes of many scientists, the Climatic Research Unit, based in Norwich, is the most expert in the world in the study of the problem.

The Unit reports overwhelming evidence of warming. The rise of 0.5°C (0.9°F) that they have detected this century represents the fastest rate of warming of the planet for thousands of years. They predict a further rise of 1.5°C (2.7°F) or more by 2030 or 2050 if the greenhouse effect becomes reality. But the Unit does not ascribe the current warming to the greenhouse effect. They say that the temperature rise experienced is a *likely* result of global warming, accompanying the burning of fossil fuels but they would stress that weather is variable and unpredictable and that other factors may be involved. The members of the Unit will not link the 'one-off' weather extremes, such as those noted in Chapter 1, with the greenhouse effect either. One of their keywords seems to be 'caution'—academic theories can only gain validity through proof, and there is not yet sufficient proof, although the likelihood is overwhelming. The second keyword would appear to be 'warning'.

The Climate Research Unit was quizzed energetically about the Great Storm of October 1987 in Britain. Some observers claim global warming would decrease the likelihood of such gales. Others claim that such severe gales and the violent storms that struck Britain in January and February 1990 could be repeated at regular intervals if greenhouse warming becomes a reality. Higher sea temperatures could result in atmospheric changes over Northwest Europe and the North Atlantic, bringing more anticyclones, and thus increasing the number of warm dry summers, but bringing colder winters than the mild conditions experienced in the late 1980s. Anticyclones could result in more easterly winds in winter, which would favour heavier and more frequent falls of snow and considerably lower temperatures. The normal British weather pattern, derived from warm westerly winds, would be effectively 'blocked out' (see p. 25) if there were to be low pressure nearby.

The ferocious storms in Northwest Europe in early 1990 were an interesting abnormality. There was nothing unusual about the strength or the duration of the very windy weather. What was abnormal was *where* that weather was experienced. Similar gales are frequent to the North of Scotland and over the seas between the Faeroes and Iceland during the winter, and often in other seasons too. The 1990 gales seem to have been further South, whereas the greenhouse warming convention implies that climatic belts in the northern hemisphere will move North. But severe gales might become less frequent in Britain if the climatic belts shift. Violent storms—of the type that visited the astonished English Home Counties in 1987—could though become more common as the path of tropical storms took on a more northerly aspect. The snag with this theory is that the tropical storms occur in the North Atlantic between June and December; they do not occur in January and February even in the Caribbean, and it has been in January and February that the greatest number of violent storms have been experienced in Europe in recent years.

CONFERENCES

In May 1990 the United Nations held a conference on the environment at Bergen in Norway, and a world conference on climate was held in Geneva, Switzerland, in the autumn of 1990 (see p. 176). These gatherings are a reflection of growing international concern over global warming, and at the Bergen conference David Trippier, then British Minister for the Environment, publicly accepted that there was definite evidence for global warming. Conference delegates heard repeated calls for rich Western countries to pay incentives to Third World nations to persuade them to curb pollution and to forego certain environmentally unfriendly industries. Attention focused on attempts to dissuade India and China from developing CFCs as refrigerants. As the non-greenhouse

---

SEVERN BARRAGE

If the burning of fossil fuels is a major source of greenhouse gases, there are 'environmentally-friendly' alternatives. In Britain, the Severn Tidal Power Group, which is funded by the Department of Energy and a group of private civil engineering companies, proposed the construction of a tidal barrage across the Severn Estuary, from Weston-super-Mare to a point halfway between Cardiff and Barry. Civil engineers Taylor Woodrow claimed that a tidal barrage would have a life expectancy of 130 years—opposed to 30 to 40 years for a conventional power station—and could generate about 7 per cent of the 1989 electricity demand of England and Wales. The barrage and similar schemes have been proposed as a 'clean' alternative to fossil fuel-burning power stations which add to the emission of the greenhouse gas carbon dioxide.

Above Alternative sources of power that do not burn fossil fuels. Luz in the Mojave Desert, California (top), has the biggest solar plant in the world. Also in California are the great wind farm at Tenachapi Pass (bottom) and the windmills at Palm Springs (opposite). (Gamma)

alternatives are expensive, both countries are looking for financial assistance.

The cost of taking measures to prevent climatic change, if the extent of the greenhouse forecasts were to prove correct, would be enormous. John Easton, an official of the US Energy Department, told delegates that the bill for the United States to take the measures that have been proposed by many conservationists would be equivalent to the entire American annual gross product.

Debate over curtailing the emission of CFCs and carbon dioxide has been acrimonious.

Seventy nations attended the Atmospheric Pollution and Climate Change Conference at Noordwijk, in the Netherlands, in the spring of 1990 and although there was general agreement that the emission of those gases which could lead to global warming should be limited, neither targets for reductions nor dates to achieve reductions were set.

Scientists meeting in Toronto, Canada, in 1988 called for the world community to meet a 50 per cent reduction in the emission of carbon dioxide, the principal greenhouse gas, by 2030. The pamphlet *Nuclear Power and the Greenhouse*

WIND POWER

The British Wind Energy Association has suggested that somewhere between one sixth and one fifth of the United Kingdom's energy requirements could be met by wind power. This is a similar fraction to that presently contributed by nuclear power.

Wind power no longer means an experiment at the alternative technology centre at Machynlleth, in the Welsh county of Powys, nor is the production of electricity by wind power in Britain likely to be confined to small generators on remote farms. The Central Electricity Generating Council is committed to the establishment of a number of wind farms, each comprising dozens of large modern 'windmills' with huge blades. One is to be built on an exposed peninsula in Cornwall, another high in the Pennines in northern England, while the third is being constructed on an area of bleak Welsh moorland.

On the face of it, wind power holds considerable promise but findings from a University of East Anglia study on the strength of the wind over the past few decades suggests that the basic resource necessary for that generation of power is a highly variable and declining commodity. The study reveals that the United Kingdom is less windy now than it was 50 years ago, a finding that will surprise all those who suffered the 'hurricane' of 1987.

The project compared wind data over a

50-year period from weather stations on or near the North Atlantic. It seems that the 1930s and 1940s were conspicuously windier than the 1960s and 1970s, with a decrease in recorded mean wind speeds along Britain's western coast of up to 15 per cent. A natural cycle in wind strength could be interpreted from these findings.

Greenhouse warming could also affect the strength of the wind in temperate latitudes and it has been suggested by the University of East Anglia study that wind speeds over Britain and Northwest Europe could drop. This stands in opposition to the theory, detailed above, that global warming will shift climatic belts to such an extent that Britain will experience tropical storms or hurricanes, although these would, of course, only occur very occasionally at any one site.

The conclusion that global warming could make Britain much less windy was drawn from the proposition that the greenhouse effect is likely to increase temperatures at the Poles to a greater degree than at the Equator. At its simplest, the world's wind pattern is an attempt to equalise air pressure between the Equator and the Poles, with air movements—or wind—rushing from the former to the latter. If the temperature difference between these extremes is lessened, the movement of air between the two would not be so great, with the greatest fall in wind speeds being in the middle latitudes.

Effect, published by the United Kingdom Atomic Energy Authority in 1990, proposed that a vigorous expansion of nuclear power by the developed nations would be one way of helping to prevent global warming. The pamphlet's authors claimed that an expansive programme of building nuclear power plants by developed nations could result in nuclear energy providing half of the world's energy requirements in 30 years' time and could reduce the emission of carbon dioxide by 50 per cent well before 2030.

Wind and sea power have also been proposed as alternative sources of energy, while efficiency in energy usage could cut a deal of CO_2 emission too.

LEAN BURN

Carbon dioxide is thought to account for about half of the potential warming by greenhouse gases. Industrial and technological processes represent a far higher proportion of the source of the gas than the burning and clearing of rain forests, but one of the causes most commonly identified—the consumption of oil, coal and gas in power stations—is responsible for only around 14 per cent of the carbon dioxide released into the atmosphere. In proportion to the energy produced, the main contributor to airborne carbon dioxide is the internal combustion engine particularly in cars, vans and lorries.

This fact has prompted motoring organisations, such as the Automobile Association in Britain, to encourage a more responsible use of the motor car to reduce the consumption of fossil fuels, and has spurred others to call for financial measures to tax 'gas-guzzling' vehicles off the roads. Scientists are developing a 'lean burn' engine for cars which could be up to 10 per cent more energy efficient. Research is continuing into engine size and the number of revolutions achieved per minute in engines as both of these elements have a direct bearing upon fuel efficiency, which, in turn, affects the degree of emissions.

The main contributor to carbon dioxide in the atmosphere is the internal combustion engine. Petrol stations may offer 'lead-free' petroleum but even that is not 'environmentally friendly'. (VSMISCS)

PLANKTON

Ice from the polar icecaps presents a fascinating climatic record of conditions some 10 000 years ago. The icecaps make an instructive model for global warming. Temperatures in that period increased substantially and carbon dioxide emissions rose—high levels of carbon dioxide are found in polar ice 10 000 years old. The ice also reveals decreasing levels of methyl sulphonic acid, a substance produced by oceanic plankton, indicating a rapid diminution in the plankton population of the oceans.

Plankton are the very smallest animals and plants, yet they have a profound effect upon the balance of carbon dioxide, the major greenhouse gas. Plankton absorb a vast quantity of carbon dioxide and can thus be said to be helping to prevent global warming. But if the oceanic temperatures were to increase, the currents would be disturbed, thus having an adverse effect upon the plankton. Warming would probably diminish the plankton population, which would mean that less carbon dioxide would be absorbed. This has been interpreted as the beginning of a vicious circle in which diminishing oceanic plankton would result in increased carbon dioxide which would increase warming which would further reduce the plankton.

The phenomenon, termed the 'plankton multiplier', can be deduced from studies of the polar icecaps and has been identified as a contributory factor that might lead to a 'runaway greenhouse effect'.

It sometimes seems, though, that greenhouse scientists propose as many 'correcting devices', as they suggest possible causes, for a 'runaway greenhouse effect'.

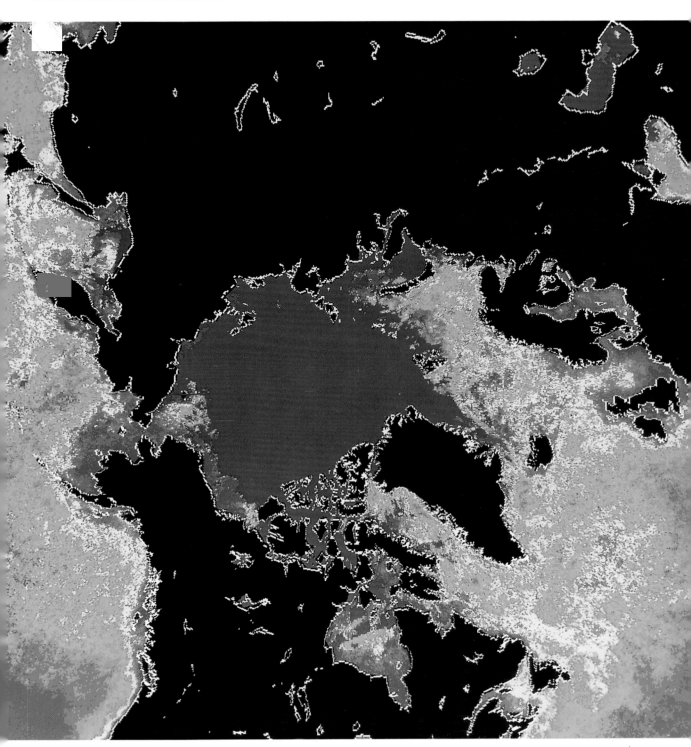

Left and above Plankton plays an important role in influencing the temperature of the oceans. The light micrograph (left) shows the diatom Pluralsegma angulator. The false-colour satellite image (above) shows the distribution of phyto-plankton in the surface water. The colours represent varying phyto-plankton densities from red (the most dense) through yellow, green and blue to violet (the least dense). Land areas are coloured black and regions where no measurements were possible are coloured grey. Highly seasonal blooms character-ize these polar waters as the long dark winters prevent photo synthesis (Science Photo Library)

REDUCTION OF GREENHOUSE EMISSIONS

The United States and Japan have both expressed an intention to reduce their emissions of carbon dioxide. The USA has given no indication of the amount of reduction planned nor the timescale for that cut. Japan announced measures late in 1990. The European Community plans to have stabilized the rate at which carbon dioxide is released into the atmosphere by the year 2000 and to have achieved real reductions in that emission within 20 years, yet in 1990 the British Department of Energy released a forecast that the United Kingdom would be releasing 37 per cent *more* carbon dioxide into the air by 2005. Germany, on the other hand, has announced that it will come close to achieving a 20 per cent reduction in carbon dioxide by 2050, even allowing for the integration of the East German economy which had not previously observed such a degree of concern for the environment.

The Netherlands has begun a compulsory 'energy audit' for its firms to encourage lower energy consumption, and France has achieved .reductions in carbon dioxide emissions, partly through a nuclear energy programme which now supplies 70 per cent of the country's electricity requirements. It has been estimated, on the other hand, that the countries of eastern Europe will take at least two decades to stabilise their emission of carbon dioxide at current levels, and it is generally recognised that the most that can be achieved in the case of most Third World countries would be a two or three per cent growth in carbon dioxide emission for the foreseeable future.

Opening the British Meteorological Office's climatic change centre at Bracknell in Berkshire, in 1990, the then British Prime Minister, Margaret Thatcher, stated that the problem of global warming had to be solved by the present generation if it were to be solved at all. Her position reflected that of many world leaders who appear to have accepted a consensus of scientific opinion that global warming is a real threat, even if its effects have not yet been felt.

Responses to that threat have varied from those who have advocated measures to prevent damage to the environment and possible climatic change to those, largely American, leaders and advisers, who believe that it would be easier and cheaper to take measures to live with the results of any climatic change than to attempt to prevent it by cutting the consumption of fossil fuels.

A panoply of industrial, technological, agricultural and environmental measures has been proposed to avert any climatic change. These include a massive programme of nuclear power station construction to cut carbon dioxide emission, national measures to conserve power at all levels (such as reducing central heating levels, using fewer luxury gadgets and so on) to reduce energy consumption, punitive taxation on vehicles and on petroleum, and massive investment in public transport to

US POSITION

The debate about the possibility—or the probability—of global warming is hotter in the United States than in most other Western countries. The official position remains that more research is needed.

At the inelegantly named White House Conference on Science and Economics Research Related to Global Change in 1990, the presidential science adviser, Dr Allan Bromley, conceded that 'if we continue to load our atmosphere with greenhouse gases we will eventually get additional warming. What we do not know is the timing, the magnitude or the rate of that increase.' However, the White House chief economic adviser remarked that 'those scientists who do believe there will be appreciable global warming over the next century are talking about an amount that would roughly translate into a change in temperature of less than moving from Washington to Atlanta.' Thus, although the concept of global warming appears to have gained acceptance, there is far from unanimity over its interpretation and—even less worldwide—over possible responses and whether it is a cause for concern.

reduce dependence on the internal combustion engine. Some observers have advocated switching power stations from coal to natural gas as the latter emits only one half of the carbon dioxide released by burning coal. Others have suggested installing chemical scrubbers to large combustion plants to cut the emission of sulphur dioxide, a major component of acid rain. (Germany has spent 25 billion D marks fitting chemical scrubbers to power stations to cut these emissions.) Longer term proposals include the use of wind, wave and solar power as environmentally friendly alternatives.

Carbon dioxide has clearly been recognised as the principal culprit but many of the possible targets for reduction in carbon dioxide emissions are challenging. It has been estimated that cuts of between 60 and 80 per cent of most greenhouse gases would be necessary to avoid any increase in the quantities of those gases in the atmosphere. Many greenhouse gases are long-lived and past and present emissions could be effective for very many years.

Measures to conserve the rain forests have been proposed and some steps have already been taken (see p. 123). Reafforestation and the planting of entirely new stands of trees are being encouraged (see p. 126). Proposals for agriculture to play a part have also been made with calls to cultivate more crops with a higher albedo value (see p. 99) and to stock far fewer cattle to reduce the emission of methane gas.

Environmentalists have put forward many ideas to help conserve the polar icecaps, mainly measures to control marine pollution and protect the plankton that absorb such large quantities of carbon dioxide.

More extreme suggestions have included proposals to create large artificial clouds to reflect incoming solar radiation to lessen warming, as it were, at source.

TOO LATE?

The scale of the problem appears daunting. It is too early to know what will happen, if global warming will bring about the climatic changes that have been forecast or even if global warming will occur to the extent that many have suggested. Some scientists have stated that the pollution of the atmosphere is irreversible and that at least some climatic change is, therefore,

inevitable. This is the view expressed by Dr Richard Warrick of the Climatic Research Unit at the University of East Anglia when he addressed the British Association in 1990. The Unit, at Norwich, is one of the leading centres for the study of climatic change, and has been at the forefront of observation and prediction of the possible effects of global warming.

Dr Warrick predicted that a rise of 1°C (1.8°F) in average global temperatures is unavoidable, given the levels of greenhouse gases currently in the atmosphere. The level of carbon dioxide in the atmosphere has risen from 315 parts per million in 1960 to 351 parts per million in 1990. It has been estimated that the pre-industrial level of carbon dioxide in the atmosphere was 280 parts per million. Computer models forecast levels of 560 parts per million by 2015. The Unit has calculated that it would take between 10 and 30 years for the full effects of the greenhouse gases already in the atmosphere to become evident and that immediate reductions could not halt the processes involved. Dr Warrick compared the greenhouse effect to the pollution of rivers by sewage and the land by chemicals, but stressed that whereas the pollution of the water and the land was eventually reversible there is no known way to reverse the current pollution of the atmosphere that he believed would lead to a catastrophic global warming. The Unit envisages an increase in average global temperatures of 4.5°C by the year 2030 unless the emission of carbon dioxide, methane, chlorofluorocarbons (CFCs) and nitrogen oxides are drastically curbed.

HEAT SINK

A natural 'heat sink' acts as a corrective device in trapping solar energy and helping to prevent overheating of the atmosphere. The heat sink is a gigantic 'trap' of energy transported deep into the oceans in the polar regions and released gradually by dispersal through the ocean currents. At present up to a third of the solar heat-energy trapped in the atmosphere (see fig. 13) is absorbed by the oceans and carried down to profound depths. The 'sinking' occurs because of the difference in the density of the salt water of the oceans and fresh water from melting ice.

The heat sink might cease—and with it temperatures on Earth might rise suddenly—if, as has been predicted, rainfall in the upper latitudes were to increase greatly. A large influx of fresh (rain) water in to the polar seas would decrease the density of the oceans' surface water and probably prevent it from sinking. This would block the redistribution of heat through the ocean and prevent the absorption by the sea of that vital one third of the radiation trapped in the atmosphere. This theory is part of the 'worst case scenario' of global warming and envisages a major, sudden climatic change as a result. Yet, although computer models support the hypothesis, it can neither be proved nor disproved—until it happens. This sink is being investigated by the Italian Terra Nova mission.

TERRA NOVA

The Italian base of Terra Nova in Antarctica is the operational headquarters of a six-year programme of research into climatic change in the southern polar continent and its adjoining seas. The research programme involves five major expeditions with a total complement of 250 scientists, technicians and others, three ships and a plane. The *Polar Queen* set sail in 1985 and has since been joined by the *Explora* and the *Cariboo* on a project organised, with Italian government funding, by the National Committee for Nuclear and Alternative Energy (better known as ENEA).

The Italians have tested Antarctic water for salinity, temperature, density and its oxygen content. They are seeking to gather data on climatic change but are keeping an open mind about how much any change may be due to human activity rather than natural causes. The Italian team feels that some ecologists have overstated their case in ascribing global warming purely to the human agent, rather than acknowledging the part that natural climatic cycles and many other factors might play.

CHAOS THEORY

Many people have assumed that climatic change is a slow and gradual business, spread over several hundred years or more. But from the fossil record, from carbon dating, from the study of rings in the trunks of trees, evidence is beginning to suggest that climatic change is altogether faster than previously imagined. Such a rapid climatic change has been proposed as one of the many theories formulated to explain the disappearance of the dinosaurs some 65 million years ago.

If the climate does make sudden jumps, with massive and unpredictable results, the reliability of our weather patterns, of forecasting and of setting up computer models to predict future climate could be called into question. But mathematics may come to the rescue.

Mathematicians refer to sudden jumps, or 'flips', of this nature as 'positive feedback', the result of which may appear random, complex and chaotic. In everyday life, a small-scale whistle in a microphone may 'flip' into an ear-piercing screech as 'positive feedback' from a loudspeaker. The 'chaos theory' holds that similar principles can be applied to climate.

This field was extensively pioneered by Edward Lorenz at the Massachusetts Institute of Technology, a quarter of a century ago. Lorenz, a most innovatory meteorologist, developed his 'chaos theory' of the complex interactions in the atmosphere by means of computer simulation. Today his theory is known as the Butterfly Effect.

Lorenz constructed a mathematical model in which the weather at any one moment is represented by a dot. Changes in that weather are then shown in the model by a series of additional dots, marking how the weather may alter from that initial state. Each dot is unique in position and is never repeated, just as identical weather conditions are never repeated. Each computer model created in this way is unique, as the changes it plots are unique, but the shape that they form is curiously always that of a butterfly!

These 'butterflies' highlight the orderly and the disorderly aspects of the mathematical calculations and indicate that a minute event could 'flip' the weather one way or another.

Edward Lorenz argued that the flap of a butterfly's wings could start a chain of events ending in a hurricane. He held that for every eventuality there was a possible opposite result. Plotted by computer model, the predictions for any climatic event form the shape of a butterfly with two wings—on one side are the likely results, on the other the 'chaotic' result of deviations.

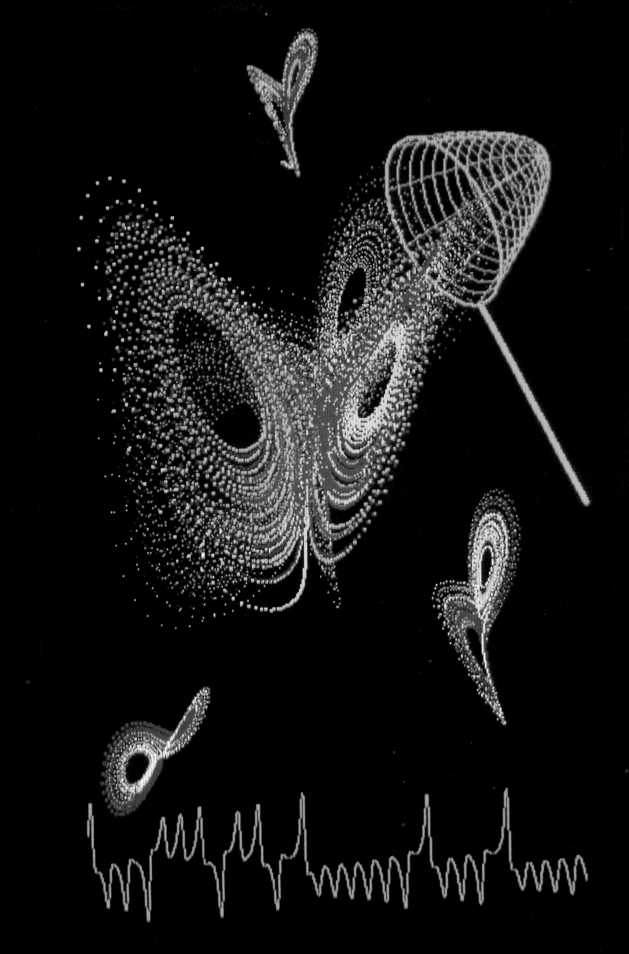

The 'butterfly' demonstrates that, beginning with one defined state, there are two possible outcomes and from a given set of climatic data two different weather systems could evolve—the two 'wings' of the butterfly.

The random nature of the 'chaos theory' indicates that a seemingly insignificant happening can trigger 'positive feedback' and through a series of countless other events lead to a momentous event elsewhere. In the classic Lorenz example, the flapping of the wings of a single butterfly—a real butterfly, not a computer simulation—can eventually trigger a hurricane, each event in the chain being a sudden 'flip'. The theory takes its name from the example of the flapping of the wings of a real butterfly—the appearance of 'butterfly wings' (more correctly referred to as the Lorenz attractor curves) on the computer simulation is a strange coincidence.

On a practical level, could the climate 'flip'? The (American) National Center for Atmospheric Research, which is based at Boulder in Colorado, has modelled the possibilities of large-scale 'flips' or switches in ocean currents as a result of global warming. They have predicted that switches in the ocean currents could occur by 2020 if positive feedback happens. This speed of change is between 300 and 600 times faster than would be predicted if the rate and quantity of greenhouse gas emission alone is taken into account.

The 'switch' most feared would be any change to the complex system of heat exchange in the oceans between the northern and southern hemispheres. In this circular system, heat from the oceans of the southern hemisphere—which warm more rapidly than those of the northern hemisphere—is circulated North, while cool currents run South, eventually warm in the Pacific and continue the cycle. A positive feedback effect could disrupt this global flow and change the entire climatic pattern of the world.

The whole is a finely balanced mechanism. Any changes, brought about for example by global warming, could upset such a balance and at any stage 'flip' the climate with incalculable results.

LONG-TERM FORECAST

Is the greenhouse effect irreversible? If warming does occur, would the resultant climatic change last indefinitely? These are some of the questions that have been faced by Dr Jean Palutikof of the Climatic Research Unit at the University of East Anglia. Her findings, through the study of climates of past periods and computer modelling, predict another Ice Age in between 50 000 and 60 000 years' time, if there is no greenhouse warming. But if substantial global warming occurs, she believes that the greenhouse effect will moderate the natural trend towards another glaciation.

Periodic glaciations are part of a natural cycle of climatic change and it has long been suspected that we are currently between Ice Ages. There is little agreement concerning the next advance of the ice. Some scientists in the 1960s and 70s proposed a glaciation as soon as the 21st century, although most now take a much longer term view. Dr Palutikof's estimate places the next glaciation further into the future than some other scientists, but a growing number agree with her that the effects of global warming could greatly moderate that glaciation. She believes that, with greenhouse warming, Britain will experience a climate very similar to that of the frozen wastes of the tundra.

The expectation of tundra conditions in Britain is a clear indication that Dr Palutikof does not expect the period of global warming caused by the greenhouse effect to last. Indeed, she points out that greenhouse warming will inevitably cease when the world's reserves of coal and oil—the fossil fuels—run out in a couple of hundred years' time. There is, however, a strong possibility that the after-effects will continue for thousands of years.

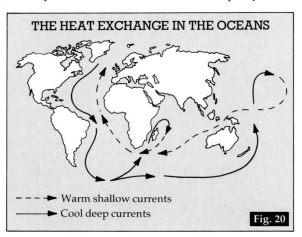

THE HEAT EXCHANGE IN THE OCEANS

- - - ▶ Warm shallow currents
——▶ Cool deep currents

Fig. 20

Jean Palutikof has built up two very long range 'weather forecasts'. She feels able to predict the type of climate that will be experienced in Britain over the next 125 000 years, both with and without the effects of greenhouse warming. Her predictions have taken into account such probable changes as a shifting of the Earth's orbit (which is possible over a period of 41 000 years) and of a likely variation in the Earth's axis (which occurs over a period of 22 000 years). Such changes as these would alter the alignment of the seas and land masses to incoming solar radiation, and bring about changes in ocean currents and the position of the climatic belts. Summaries of Jean Palutikof's two possible climatic scenarios for Northwest Europe follow.

NW Europe without global warming

Present day to c. 4000

The present temperate climate would be likely to continue in Britain with only small-scale changes associated with such phenomena as sunspot activity.

c. 4000–c. 20 000

Temperatures would gradually fall and the climate would cool into something approaching the boreal climate, presently experienced in most of Norway. Summers would be less warm. Winters would be longer and harsher, and precipitation would generally increase, much of it snow.

c. 20 000–c. 25 000

A much colder periglacial period would follow during which the climate would resemble that now experienced in the tundra of Siberia.

c. 25 000–c. 42 000

Temperatures would rise again to another boreal period, similar to that predicted for the epoch from c. 4000 to c. 20 000.

c. 42 000–c. 52 000

Another sharp fall in temperatures would bring a second short periglacial period.

c. 52 000–c. 62 000

The next Ice Age is predicted at this stage of Dr Palutikof's model. Thick sheets of ice would cover Britain and the sea level would fall by around 60 m (nearly 200 ft).

c. 62 000–c. 65 000

Another very short periglacial epoch would follow the Ice Age as natural warming took place.

c. 65 000–c. 100 000

Dr Palutikof predicts a long period of remarkable climatic stability during which a boreal weather pattern would be established. Conditions would resemble those forecast for the period between c. 4000 and c. 20 000.

c. 100 000–c. 115 000

A further cooling into another tundra epoch is predicted.

c. 115 000–c. 125 000

After the traumatic changes that took Britain through colder, wetter weather, a harsh tundra climate and another Ice Age, the climatic model brings us back to something resembling our present variable temperate climate.

NW Europe with global warming

Present day to c. 3000

Britain would experience a subtropical climate very similar to the conditions in the Mediterranean countries today. Mild relatively rainy winters would alternate with hot dry summers, but overall precipitation would decrease.

c. 3000–c. 25 000

Temperatures would gradually fall as the long-term effects of global warming wore off, and the climate would return to resemble the conditions that are currently experienced in the British Isles.

c. 25 000–c. 50 000

The natural cooling that Dr Palutikof has forecast appears in her second model as a long period of boreal climate with temperatures and rainfall very much as they are predicted in the first model.

c. 50 000–c. 65 000

The effects of global warming are forecast to be prolonged. In the second model, instead of another Ice Age Dr Palutikof predicts an epoch of harsh tundra conditions.

c. 65 000–c. 100 000

A long period of boreal climate is predicted, exactly as in the first model.

From the year 100 000, the climatic pattern predicted in this model is exactly the same as the first forecast. The long-term effects of greenhouse warming have ceased to be a factor, and the climatic change wrought by man's economic activities is seen to be geologically short-term.

11 Current research

A contributory factor in global warming could be the activities of the little understood Pacific Ocean current, known as El Niño. Studies by the American National Oceanic and Atmospheric Administration suggest that variations in the weather caused by the warm El Niño current and by other similar currents may have been underestimated. The Administration believes that by ignoring the (temporary) effects of these currents, scientists may have seriously overestimated the rate at which global temperatures are increasing.

The current was discovered by fishermen working out of Spanish ports on the Pacific in the 17th century. They named it El Niño—meaning the Christ Child—because, although it occurs irregularly, in the years in which it does happen, it flows around Christmas.

This mysterious current begins in the Pacific between Papua New Guinea and Micronesia and flows across the vast ocean towards Peru. El Niño occurs every three to eight years when the sea North of New Guinea warms, reaching temperatures up to 30°C (86°F). The problem of El Niño is complex. It is difficult to know what is cause and what is effect.

EVIDENCE FOR EL NIÑO

Meteorological interest in El Niño and other ocean currents is developing rapidly. The British Meteorological Office has researched sea surface temperatures and any possible relationship that might exist between them and weather patterns. By 1989 they had established a definite correlation between the temperature of the oceans and the rainfall totals falling—or often *not* falling—in the Sahel of West Africa.

The Meteorological Office team studied weather data going back several decades and found that periods of drought or abnormally low precipitation in the Sahel, correlated with periods during which there were considerable differences between the surface sea temperatures (often referred to as SSTs) in the southern hemisphere and the northern hemisphere in spring.

Previous page Modern meteorology uses computer modelling to forecast conditions such as drought, damaging frosts and hurricanes. (SEFA)

It has been demonstrated that unusually warm seas in the southern hemisphere in spring are unerringly followed by drought in the summer in the Sahel. The reason suggested for this is that if the South Atlantic warms, pressure will be reduced there as a result. The creation of this weaker high pressure causes a decline in the Southwesterly winds that would otherwise bring rain to West Africa and thus less rain is received by the Sahel. The relative timing of the phenomena seems consistent with higher South Atlantic temperatures in March or April being followed by drought in Niger, Mali and Burkina Faso in July and August. A clear relationship also seems to exist between June SSTs and rainfall.

Ocean temperatures in three oceans—Atlantic, Pacific and Indian—seems to be correlated with the rainfall totals recorded in West Africa. Research is now under way to discover if there is such a definite link between sea surface temperatures in all the oceans and in seasonal rainfall totals in other areas, such as Brazil and Australia. Inevitably, thoughts must turn to El Niño, the most dramatic of the ocean currents. When El Niño flows there are exceptionally warm sea surface temperatures, beginning in the eastern Pacific between Papua New Guinea and Micronesia. Although no correlation has yet been found, the growing consensus is that it does seem probable that ocean currents, especially the powerful El Niño, have a major influence upon our weather.

LA NIÑA

Interaction between the ocean and the atmosphere remains the most popular theory for the cause of El Niño, but growing numbers of scientists are suggesting that the warm current is triggered by volcanic activity either on or just below the sea bed.

A reverse situation, tentatively christened La Niña, also exists. This involves a cold Pacific current being diverted much further North than usual and so disrupting the normal weather patterns in the northern hemisphere too. Both El Niño and La Niña are thought to bring major short-term changes in the weather patterns, shifting air masses to the North or to the South of their customary locations. The side effects of La Niña ('the girl child') are said

OPERATION VIVALDI

The role of the oceans in both determining climate and in influencing possible climatic changes is still as mysterious as the deepest reaches of the oceans themselves. Oceanic physics is a 'growth science', and, if the possible repercussions of global warming upon the seas are to be understood in time, a great deal of research has to be undertaken.

A major oceanic physics research programme is being conducted by the World Ocean Circulation Experiment, whose activities are co-ordinated by the University of Southampton. Funded by 44 countries, the programme aims to collect data on the immense cycle of movements of warm and cold water in the world's oceans.

The National Environment Research Council in Britain is funding important independent studies into the polar seas of the North Atlantic. Called Operation Vivaldi, the project will involve extensive surveys of the far North Atlantic by a research ship. A 'sea soar', towed by the vessel, will measure salinity, temperature, depth and other factors as part of a study of oceanic salinity to gain a clearer understanding of the workings of the 'heat sink' (see p. 162).

Other aspects of Operation Vivaldi will include studies of the melting north polar icecap and of levels of absorption of carbon dioxide by plankton. These projects will use data collected by small robot submarines capable of making extensive journeys under the Arctic ice.

to include greatly increased rainfall in Australia (which would seem to be confirmed by the events of 1989–90), less rainfall in South America (which does seem to be the case in part of Brazil) and, at a greater distance, a dislocation of the path of cyclones in parts of the northern hemisphere. Some people have attempted to explain the stormy weather experienced in much of Europe early in 1990 in this manner, pointing out that there was nothing unusual about the ferocity of the storms that hit Northwest Europe except that those conditions are normally experienced in slightly higher latitudes at that time of year.

Whatever is said about El Niño, or about La Niña (whose existence some observers deny), is controversial. There is very little agreement about the nature or the effects of these currents. But they are an additional and probably an important factor in the 'weather machine' whose complexity we are only just beginning to grasp.

QUASI-CYCLES

The atmosphere of planet Earth is an interactive system. The air masses and weather conditions in one zone cannot be considered in isolation as changes in the part may influence the whole. Climatic variations in one region can be caused by, or have an effect upon, the weather in other regions.

Particular attention is being focused upon how variations in tropical climates may have a knock-on effect in areas outside the tropics. Tropical climates have traditionally been regarded as unchanging, yet recent observations have revealed a quasi-cyclic pattern of variation in these climates over periods of between 40 and 50 days. These irregular or quasi-cycles were initially recorded by satellites in the early 1970s, and were, at first, thought to be of little importance, but research in the last few years has shown there is a link between these quasi-cycles and certain short-term climatic patterns in areas outside the tropics.

Every 40 to 50 days, warming intensifies in the Indian Ocean, leading to rising air masses and the formation of impressive banks of cloud. As the cloudiness increases, these moist air masses move East across the Indian Ocean and into the Pacific for a period of one month to six weeks. There has been no record yet of these swirls of cloud reaching South America, but they are believed to influence weather over a very wide range. There have been (unconvincing) attempts to connect these cycles with heating in the Pacific Ocean, off Papua New Guinea which triggers the current El Niño. However, the cycles are so dissimilar in length—40 to 50 days in the case of quasi-cycles and several years in the case of El Niño—that any connection is improbable, although it is possible that the effect of the quasi-cycles is cumulative and that the additional heat from one quasi-cycle might be enough to trigger the beginning of an El Niño period.

Of far greater importance is the correlation that has been observed between quasi-cycles in the Indian Ocean and blocking in temperate latitudes. The nature of that link is as uncertain as the reasons why these irregular periodic fluctuations occur in the first place. Research into the causes of the quasi-cycles is being conducted, but early answers cannot be expected. In the meantime, the effects of quasi-cycles add one extra problematic dimension to our understanding of climatic variation.

CURRENT THEORY

The warmth of the sea triggers El Niño, which—in a way not yet fully understood— sometimes interacts with the atmosphere. It used to be thought that the only major consequence of El Niño was occasional disruption of the more usual currents and sea temperatures off the Northwest coast of South America. This affected the feeding habits of fish and resulted in small catches and economic problems for the fishermen of Peru every few years.

Following studies undertaken by British scientists on the Royal Research Ship, the *Charles Darwin*, it was realized that El Niño was responsible for climatic variations in a number of parts of the world. The current can interact with the atmosphere to reduce the seasonal monsoon air flows reaching India, thus causing drought in the subcontinent. The winds and the ocean somehow work in concert, depriving South Asia of monsoonal rain. Disruption on this scale has a knock-on effect in other parts of the globe.

Six of the warmest years on record were in the 1980s, and they can be correlated with the flow of El Niño; for example, El Niño happened in 1982–83 and this was followed imme-

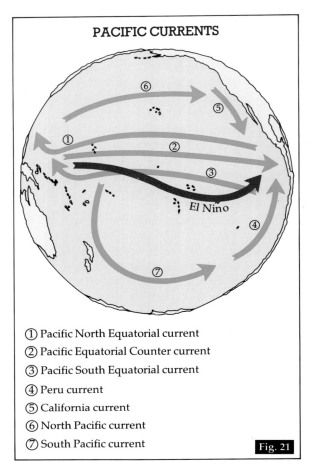

PACIFIC CURRENTS

El Niño

① Pacific North Equatorial current

② Pacific Equatorial Counter current

③ Pacific South Equatorial current

④ Peru current

⑤ California current

⑥ North Pacific current

⑦ South Pacific current

Fig. 21

diately by the warm weather of 1983. Similarly, El Niño flowed in 1986–87 and was followed by the especially warm years of 1987, 1988, 1989 and 1990.

El Niño is now thought to be linked with regular variations of wind and rainfall in the tropics, and it has been shown to correlate with drought in Australia. Its effects on other parts of the world, though, are not thought to be so profound. Nevertheless, the current is known to add indirectly to the temperature of the land surface.

One of the difficulties in ascribing all the recent spectacular climatic variations to global warming alone is that the proven amount of temperature increase is probably not sufficient to have all the results with which it is blamed. The world's surface temperature in 1989 was calculated to have been only 0.23°C (less than 0.45°F) warmer than it was in the period 1951–80. The temperature of the land has, therefore, increased but that increase is too small to have produced many side effects.

Perhaps the warming of the sea by El Niño, and possibly by other currents of a similar nature, have had a greater influence than sometimes allowed. It is possible that the climatic variations that we have experienced in the past decade are due not just to the greenhouse effect but also to other more temporary phenomena.

The mass of facts, evidence, theories and opinions is confusing and often contradictory. Many scientists are convinced that global warming is a reality and that it is caused by the emission of greenhouse gases, but in August 1990 a report was presented to the conference of the British Association, meeting in Swansea, that jolted some opinions. The Scott Polar Research Institute revealed that, far from retreating, the icecaps of southern Greenland were thickening at a rate of 20 cm (8 in) a year. This freezing, which has been shown to apply to nearly half the Greenland icecaps, is the equivalent of *lowering* the sea level of half a millimetre (0.02 in) a year. This is, of course, an insignificant amount but it is a very different story to the common predictions of a rise in sea level of a third of a metre by the year 2030. (This thickening of the Greenland cap may, however, be in line with the greenhouse theory. Warmer temperatures and great cyclonic activity will produce more precipitation which—in high latitudes or areas of high relief—will fall as snow and so tend to accumulate. Only later as warming developed would melting dominate.)

The truth is that we cannot say what is happening—yet. There is every possibility that we may not know until warming has—or has not—happened in 40 or 50 years time. Some scientists have rejected the greenhouse theory and are ascribing the increase in temperature that has happened this century to cyclic solar activity.

The most celebrated 'convert' is Professor James Hansen of NASA who in 1988 had gone on record as saying that he was '99 per cent certain' that greenhouse gases were causing drought.

At the same time, 'worst case' scenarios have gradually become less horrific over the last few years. Some estimates in 1980 forecast a rise in sea level of up to 8 m (26 ft) by 2050; current predictions average about one third of a metre or less. Similarly, in 1987 estimates of a start-ling temperature increase of 5°C (9°F) by 2050 were not uncommon; in 1990 estimates of an increase of about 2°C (3.6°F) were being made. Reports presented to the 1990 World Climate Conference in Geneva suggest even lower figures—a rise in mean temperatures by 1°C (1.9°F) by 2025 and by 3°C (5.7°F) by 2100 if the emission of greenhouse gases continues unabated. The suggested rise in sea level is also less than previously forecast—by 20 cm (8 in) by 2030 and by 65 cm (26 in) by 2100. Yet even if the predictions seem to be becoming less drastic, it is still difficult to know what to believe or accept.

ORBITS

A store of data on the climates of past periods awaits interpretation in tree rings of species such as the bristlecone pine (*Pinus aristata*). A native of the Rocky Mountains in the United States, the bristlecone pine is thought to be the longest lived of any existing tree species. Recent research on tree rings in northern Sweden suggests that the extent and the effects of the medieval 'Little Ice Age' were not nearly so severe as had been once thought, at least in Sweden.

Improvements in carbon 14 dating, made possible since the publication in 1990 of research conducted at Columbia University, New York, make tree ring research beyond 6000 BC more accurate, and a combination of carbon dating and tree ring studies is being employed in a variety of research projects. These include studies into rises and falls in the level of the oceans, into past periods of global warming (especially the warming some 10 000 years ago which was initially identified through the fossil record) and into past advances and retreats of the icecaps. These projects are largely uncoordinated and are being undertaken at universities and institutes in over two dozen countries. It is too soon to expect conclusions from these programmes, but there is a hint of a common thread of enquiry appearing in some of them. The climatic changes under investigation appear to be cyclic and to be connected in some way not yet understood with the Earth's orbit, which some scientists are beginning to believe acts as a 'trigger'.

GRAVITY

One of the most interesting theories of the causes of natural cycles in climate has been proposed by Dr Bruce Denning, who was professor of ocean engineering at the University of Newcastle-upon-Tyne in Northeast England. Dr Denning suggests that cyclic heating and cooling of the planet is caused by gravity. He believes that the Earth is pulled towards the Sun more strongly at certain periods than at others and that this gravitational pull has a profound effect upon climate.

If this were to be true, it could be possible to calculate those changes in temperature in the past that have been due to natural warming, and to define what proportion of those temperature increases currently being experienced are the result of natural processes and what proportion are due to human activity.

Evidence of climatic conditions in the past is held as isotopes of oxygen in sedimentary deposits on the seabed. Study of these deposits could reveal the timescale of natural climatic cycles over millions of years. Dr Denning has used such findings to construct a computer model of natural climatic change up until the 21st century. His findings indicate a rise of 0.75°C as a result of natural effects. If this is added to the usual forecast of a rise in temperature owing to the greenhouse effect of 1°C by the year 2000, Dr Denning predicts warming at double the rate we have almost come to expect. He is concerned that scientists should not mistake a faster rate—that is greenhouse warming plus natural warming—as proof of a runaway greenhouse effect.

Dr Denning developed his theory while engaged in predicting weather patterns in the North Sea for oceanic engineering projects for oil companies. He discovered that when energy hits the surface of the sea, a change in the balance between oxygen isotopes 16 and 18 occurs. The isotopes rise with evaporated water, fall in raindrops and are eventually deposited within the layers of sedimentary rocks forming on the seabed. It is believed that an average gap of about two years separates the moment of initial impact when the isotope balance is affected and the moment when those isotopes are laid down in sedimentary deposits. In the ratio between isotopes 16 and 18, these layers contain a record of the sea tem-

peratures two years previously.

Dr Denning's theory of a gravitational impact upon climatic change and a correlated weather pattern from the past have not gained widespread acceptance, although he claims that his model successfully predicts the warming of the late 1980s. It has nevertheless not been disproved as an explanation of the cyclic nature of climate.

SUNSPOTS

In a survey of 1500 experts in climatology carried out in 1990 for the magazine *Nature*, 71 per cent of those questioned stated that they believed that the temperature rises experienced over the last 100 years are within the natural range of fluctuation. It seems, therefore, that many scientists look for a natural cause for the warming, and to many the obvious cause is solar activity.

It is easy to accept the blazing heat of the Sun as a constant. The heat of the Sun may appear to us as reliable and unchanging, particularly while we enjoy its warmth on a pleasant summer's day. But the Sun is a restless, dynamic star. Its radiation of heat varies, and it is that variation that could have a much greater effect on our climate than the greenhouse gases.

Variations in the Sun's radiation are nothing new. The planet Earth depends upon the Sun for its very existence and solar radiation is the single major determinant of climate (see p. 79). It is, therefore, understandable that conditions on the Sun and conditions on Earth are intimately connected. Over the centuries variations on the Sun have been felt in terms of increased or reduced radiation received on Earth, and there is no reason to suppose that future variations in conditions on the Sun will not be reflected on our planet. In theory, it

Right and over page This black and white computer-enhanced photograph of a giant cluster of sunspots was taken on 1 September 1989, a few months before the maximum in the eleven-year cycle of solar activity. The false-colour image of the Sun (over) shows sunspots (the green dots), active regions (the large red areas) and filaments (the green lines). Sunspots—pockets of cool dark gas—occur at magnetically active regions, often in groups. (Nigel W. Scott/Science Photo Library)

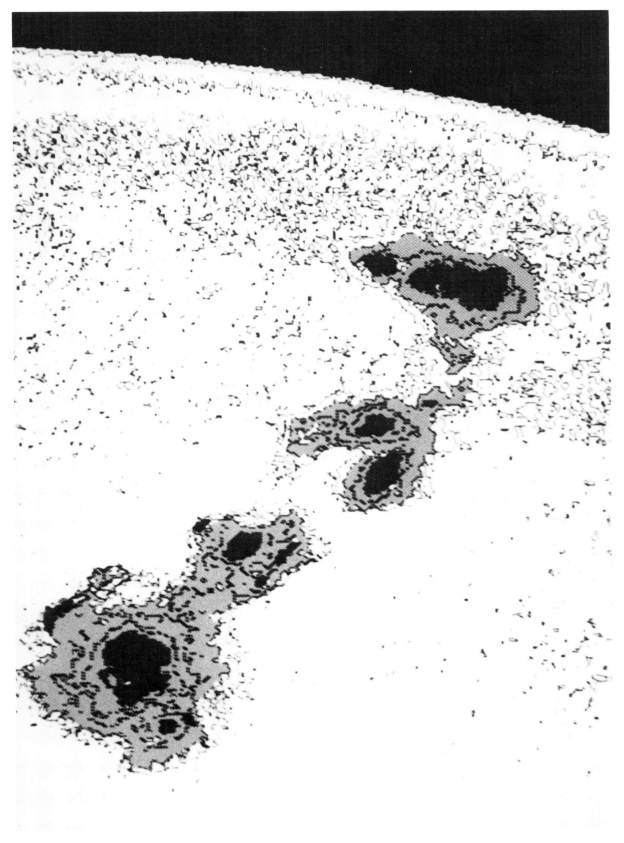

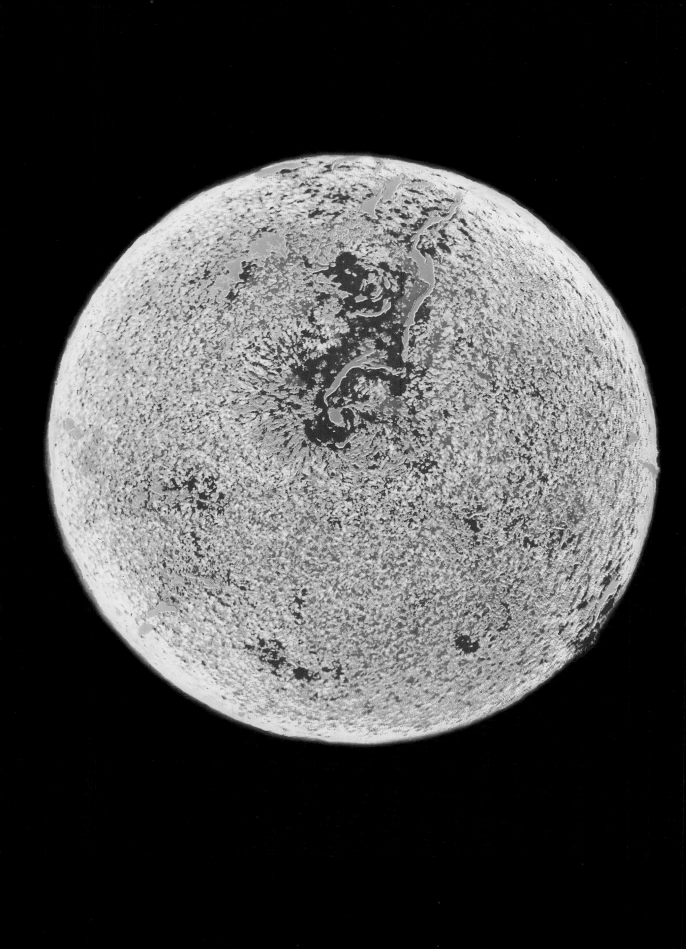

should be possible to predict future changes in the Earth's climate, if we can predict future changes in conditions on the Sun.

This field of research seems to have attracted more attention in the United States—where the very concept of greenhouse warming is hotly disputed—than in Britain, although British scientists have played their part in the study of the most obvious changing element on the surface of the Sun—sunspots.

Sunspots have been observed since at least the 4th century BC, but the ancient civilisations did not realise what they were. Until Galileo (1564–1642), it was thought that sunspots were the shadows of heavenly bodies passing between the Sun and the Earth.

Sunspots are dark markings on the Sun's photosphere—that is the boundary between the Sun's dense interior and its transparent atmosphere. Each sunspot is typically 2000–3000 km (1200–1900 miles) across and is darkest at its centre. Sunspots seldom occur individually, but rather in groups of two, three or more, and tend to be short-lived, forming and disappearing again within the span of two to three weeks. A minority may last longer. Each sunspot is a centre of intense magnetism, which is highly localised. The formation of these centres occurs in cycles.

The existence of a cycle in variations on the Sun was first suspected by a German amateur astronomer, Heinrich Schwabe, in the 1840s. He observed that the number of dark spots on the Sun's surface seemed to vary in a remarkably regular pattern of about ten years. Rudolf Wolf—working at Zürich Observatory in Switzerland in the 1850s—took up Schwabe's theories and, based on daily observations sent to him by a number of astronomers working in several European countries, Wolf identified a sunspot cycle lasting just over 11 years.

A study of the records of solar observations over a period of 150 years confirmed Wolf's theories. Taking into account the findings of past astronomers, Wolf was able to identify 13 sunspot cycles between the years 1610 and 1848. Each cycle was a period of about 11 years during which conditions on the sun altered significantly. Of greater importance to us is that those sunspot variations were reflected by variations in the climate here on Earth.

Earlier observers had recognized variations in sunspots. The notable British astronomer Sir William Herschel had correctly suggested that the Sun is at its brightest when it has most sunspots. He even went on to attempt to correlate the number of sunspots and the price of wheat, claiming that sunspot activity increased the amount of sunshine which, in turn, would provide a better harvest.

The number of sunspots observed has varied dramatically over the decades. Between 1645 and 1715 very few sunspots were observed by astronomers. This was a period of diminished solar activity. The British solar scientist Walter Maunder was the first to notice this period of few sunspots and reduced radiation from the Sun. It became known as the Maunder minimum.

Until recent years the Maunder minimum was not taken seriously by many astronomers and solar scientists. It was suggested that 17th-century astronomers had weak telescopes and would not be able to see the sunspots even if they were there. Some astronomers put forward the possibility that their counterparts in the past were somewhat lax in the regularity and accuracy of their observations. For whatever reason, their competence was questioned. However, there was a lack of sunspots, a change in the level of solar radiation, in the 17th and 18th centuries.

It is now known that the Maunder minimum coincided with the most severe period of that particularly cold time known as the Little Ice Age and which lasted until the early 19th century (see p. 60).

It has been estimated that the Little Ice Age experienced a drop in temperature large enough to correlate with a decrease in solar radiation of as much as 0.5 per cent, enough to have a real effect on our climate.

The Maunder minimum was not the only significant low, or high, in the number of sunspots that has been observed. The number of sunspots varies dramatically from cycle to cycle. At the beginning of the 19th century only 45 sunspots were observed. In the middle of the 1950s some 190 sunspots were recorded, and, as we have seen, temperatures on Earth in the mid-1950s were notably higher than they were at the start of the 19th century. There is, it seems, a connection between the number of sunspots and the temperatures experienced on Earth, although the link is still debated. Many would argue that if it were so simple, the connection would have been established years ago.

The 11-year cycle of sunspots is only one change that has been noted in conditions on the Sun. That cycle is now known to be one half of a regular 22-year cycle in the Sun's magnetism. Great progress was made in the 1970s and 1980s in understanding the variations in the Sun's output of light, now known as the solar constant. The solar constant varies even over a relatively short period of time. Readings from satellites show that the solar constant may vary by as much as nearly 0.25 per cent over a few weeks. Such variations are caused by dark sunspots passing across the face of the Sun as that great star rotates. These sudden short-lived variations have only been observed since the 1980s, as observations at the Earth's surface were marked by atmospheric effects.

Since 1987 a marked increase in solar activity has been noticed. Indeed, solar activity is currently at the highest level ever observed by man. It is perhaps significant that this large increase in solar activity correlates with the record-breaking hot summers and mild winters of the late 1980s and 1990.

The connection between solar activity and our weather has been proved to exist. In 1987 the German scientist Karin Labitzke proved the correlation of solar cycles over 40 years with warmer than average winters in Europe and North America. The pattern she observed correctly predicted the remarkably warm and mild winter of 1988–89 and the hot summers of 1989 and 1990.

Within a decade the solar cycle will have turned—the peak is in 1991—and solar activity will have decreased. The natural trend will be towards cooling, colder winters, less scorching summers and more reliable rainfall. It will be interesting to see whether the natural cycle, or a man-made climatic variation, will predominate by the year 2000.

Professor Peter Foukal, of the Kitt Peak Observatory in Arizona (USA), believes that next century will be one of conflict between man-made global warming and the cooler effects of a 'quieter' Sun. If there is to be a contest, it is impossible to predict which will win.

Professor Labitzke, at the University of Berlin, emphasised that those who predict global warming on the basis of man-made pollution by carbon dioxide need to be very careful. She maintains that a succession of mild winters do not necessarily have anything to do with the greenhouse effect.

THE ECONOMIC ARGUMENT

In the autumn of 1990 a new element entered the debate on global warming. Having studied the measures that might have to be taken in an attempt to combat the possible effects of global warming, or endeavouring to prevent warming in the first place, a number of economists have reached the conclusion that the 'economic argument' is against taking any such measures.

They argue that the cost—in terms of money, employment, life style and standards of living—would be so great that the finance invloved would be too large to be considered. For example, to prevent global warming, the transport habits and policies of the countries of the Western world would have to be radically revised. Emissions from car exhausts are an important contributor to greenhouse gases. The 'economic argument' runs that the expense of changing transport policies would be greater than the cost of coping with climatic changes in the middle of the next century. It is argued, for example, that most of the land threatened with inundation, owing to any rise in sea level, is not of great economic importance. Much of the threatened land comprises small island groups and coastal swamps. There are, of course, major cities and heavily-populated areas under threat too, but any rise in sea level would be gradual and could be countered by adequate sea defences.

It has also been argued that the changes in agriculture that could result from global warming would be an 'economic plus'. Increased carbon dioxide in the atmosphere could be expected to make plants grow more rapidly—just as it does in greenhouses. The 'economic argument' is that increased temperatures will lead to increased production of crops. Some areas would benefit from global warming. Forecasts of maize and soya beans growing in southern England, of vineyards in the English Midlands and commercial soft fruits in northern Scotland have received a great deal of attention. So have prophecies of wheat growing in the north of Russia, of Japan becoming self sufficient in rice, of cereals ripening in Iceland and parts of Australia becoming temperate farming regions. But the changes in temperature that would make such changes possible would be likely to be accompanied by significant alterations in rainfall patterns worldwide.

Many of the present semi-arid regions would be subject to erosion and desertification if their rainfall totals fell sufficiently. The major wheat lands of the world—Canada, the USA and western Europe—would all be likely to receive less rainfall in a greenhouse world and thus crop yields would fall. And if the USSR were to benefit from being able to grow cereals further to the North, it would suffer at the same time as the important grain-growing districts of the Ukraine became less fruitful because of declining rainfall.

The likely effects of global warming upon agriculture would be a general reduction in food production, an increase in food prices and a risk of famine to the one billion people living in semi-arid regions. The economic and the human costs of adapting to such challenges

would be great, but the 'economic argument' persists with a growing number of adherents who hold that it would be cheaper and easier to cope with effects of global warming than to attempt to prevent them.

WARNING

At the end of October 1990, scientists, civil servants and government ministers from 130 countries met in Geneva, Switzerland, for the Second World Climate Conference. At the time, the economic argument (see above) was beginning to attract support, but scientists at the conference warned the world's governments not to underestimate the dangers of global warming. They urged that the world's annual emissions of carbon dioxide should be cut by one or two per cent per annum globally, beginning now. If this rate were to be adhered

A Calpak BP solar heater displayed at a selling point in Athens, Greece. Energy conservation and the need for alternative sources of power has become accepted almost universally. (British Petroleum Company)

The Second World Climate Conference was held in Geneva in November 1990 at a time when the economic arguments against attempting to halt the greenhouse effect were gaining some currency. (Gamma)

to, by the middle of the 21st century, carbon dioxide concentrations would stand at 50 per cent above pre-industrial concentrations.

Such a rate would be, the scientists maintained, 'technically feasible', and they indicated that 'cost effective opportunities exist to reduce carbon dioxide emissions in all countries'. They believed that many industrialized countries could cut fossil fuel emissions by at least a fifth by the year 2005.

The Conference attempted to gain agreement for a global reduction in carbon dioxide emissions of 20 per cent by 2005, but these proposals were blocked by Canada, Saudi Arabia, the Soviet Union and the United States which is likely to increase its output of carbon dioxide by 15 per cent over the next 20 years.

The scientists meeting at Geneva were united by worries concerning decreased funding for climatic research. The World Ocean Circulation Experiment, for example, has had to be redesigned and extended over a longer period with fewer observations because of a shortfall of funding of about 15 per cent. The

climatic researchers were almost united in agreeing that natural ecosystems can cope with an overall temperature rise of 1°C (1.8°F) without any major disruption. A rate of temperature increase of 0.1°C (0.2°F) per decade could also be 'absorbed'. It is also thought that vulnerable low-lying coastal districts could adjust to a rise in sea level of about 2 cm (0.8 in) a decade.

The Geneva Conference asked governments to agree that:
—the emission of greenhouse gases be stabilized and thereafter reduced;
—long-term commitments be made to climatic research;
—more efficient use be made of energy resources;
—reafforestation should take place at a rate of 12 000 000 ha per annum.

GLOBAL PROBLEM

Greenhouse warming is a global problem. As such it can only be confronted on a global scale. Any one government cannot by itself tackle the issue. In 1990 some governments had not yet been convinced that the greenhouse effect existed, and it would be unrealistic to expect them to take measures to tackle a problem whose existence they questioned.

We are dealing in theories and forecasts. Global warming has been suspected for less than a quarter of a century, and it is too soon to know exactly what is happening. Scientists cannot agree about how quickly warming will occur. Nor can they tell us with any certainty by how much temperatures may increase. We will only have those answers once it has happened and then it may be too late to avoid the serious consequences of a man-made climatic change.

As long as the scale of the problem cannot be clearly defined, it is unlikely that the costly measures necessary to prevent the more extreme consequences that have been predicted will be taken.

We have all become aware that there is a problem and that we cannot take our climate for granted. The unusual weather conditions experienced over the past few years have increased our awareness of the issues, but it is improbable that the greenhouse effect has by itself caused the prolonged droughts, the mild winters and the ferocious storms. There has been an increase in global temperature of between 0.3°C and 0.7°C since the start of the 20th century. It is doubtful that such a small increase alone could have been responsible for the dramatic climatic variations of the past 10 years. Solar activity, ocean currents and other phenomena, such as blocking, may have played a significant role. Increased attention is being paid to theories concerning the cyclical nature of climate, and there is very considerable evidence that—owing to increased sunspot activity—we are currently in the warm period of a natural cycle.

We have probably not experienced the greenhouse effect yet, but the coincidence of a natural phenomenon with the discovery of the possible climatic results of human activity could be viewed as a timely warning.

Glossary

albedo In climatic terms, the reflecting of incoming solar light into space off clouds, dust, the atmosphere, ice, water and the land surface etc. Clouds, ice and snow have a greater ability to reflect and are therefore said to have a high albedo.

altocumulus Known as Ac. The base of altocumulus clouds varies between 2000 m (6500 ft) and 6000 m (20 000 ft). They may be white or grey and can occur in patches or sheets. Sometimes altocumulus clouds occur in 'rolls' but because of the diversity of their possible shapes, these are the most difficult clouds to identify, although a globular character is common.

altostratus Known as As. These grey or grey-blue sheets or layers of cloud occur at the same heights as the altocumulus. Altostratus sheets may totally obscure the Sun upon the approach of a warm front when they follow the more broken cirrostratus clouds.

Antarctic The area around the South Pole, usually the area South of 66° 32′ S.

anti-cyclone An area of high pressure containing, usually, warm sinking air. Winds flow out from areas of high pressure in which the air does not rise enough vertically for much rain-bearing cloud to form. Thus, rainfall is not common in areas of high pressure. In anti-cyclones, winds blow clockwise in the northern hemisphere and anticlockwise in the southern hemisphere. High pressure areas tend to bring warm, dry sunny weather to Europe in summer and cold dry weather to the same continent during the winter.

Arctic The area around the North Pole, usually the area North of latitude 66° 32′ N.

aridity index The aridity index is a measure of the *effectiveness* of the precipitation that is received. It could be described as a measure of evaporation. Thus, Death Valley in California has an aridity index of seven which means that the Sun would evaporate seven times the average rainfall it actually receives. On the other hand, some parts of the Sahara have an aridity index of 200, meaning that the Sun would be capable of evaporating 200 times the rainfall received.

atmospheric pressure The pressure exerted by the atmosphere—usually measured in millibars. Atmospheric pressure varies and decreases with height. At sea level it is, on average, recorded as 1013.2 millibars.

average In climatic terms, the average experience of climatic elements—rainfall, temperature etc—over a period of 30 years, currently taken to be 1951–80.

backing A change of wind direction in an anticlockwise direction.

bar A unit of pressure, equal to 750 mm of mercury at 0°C.

barometer A device for measuring atmospheric pressure.

barometric gradient If the isobars on a weather map appear close together, the pressure gradient or barometric gradient is said to be steep and the winds will be strong.

Beaufort Scale The scale for measuring the speed of wind in a series of numbers, 1 to 12. The number on the scale can be calculated by easily observable effects such as tree movements and the degree of damage caused. The scale was devised by the English Admiral Sir Francis Beaufort in 1805 (see p. 31).

blizzard A storm of powdery snow or ice—carried by a high wind—during which visibility is limited.

blocking The lingering presence of large anticyclones or highs that 'block' or stop the normal passage of air masses over an area.

butterfly effect The theory that the flapping of the wings of a single butterfly can eventually trigger a hurricane, each event in the chain being a sudden 'flip'.

calm No perceptible air movements.

chaos theory The theory that climate can make sudden 'jumps', with massive and unpredictable results.

cirrocumulus Known as Cc. Cirrocumulus clouds appear at the same heights and have the same base temperatures as cirrus and cirrostratus clouds. They form a white 'veil' that may cover part or all of the sky. Cirrostratus may take a 'halo' shape around the Sun or the Moon. Like the other high clouds they are formed of ice crystals.

cirrostratus Known as Cs. The base of cirrostratus clouds vary between 5000 m (16 500 ft) and 13 700 m (45 000 ft). Their base temperatures are the same as those of cirrus clouds. Cirrostratus clouds are thin white sheets or layers of cloud which normally appear without any internal markings. They may take the appearance of 'ripples' and can appear to be arranged in a regular pattern. Cirrostratus clouds appear after cirrus clouds upon the approach of a warm front.

cirrus Known as Ci. These clouds occur above between 5000 m (16 500 ft) and 13 700 m (45 000 ft). Their temperature at base level ranges from $-20°C$ to $-60°C$ ($-29°F$ to $-76°F$). Cirrus clouds are white, detached delicate clouds that occur in thin bands or filaments and sometimes in patches. They often have a silky sheen. Some have a 'hooked' appearance, hence their name of 'mare's tails'. The highest of clouds, cirrus frequently occur at altitudes of 8200 m (27 000 ft). They may be seen high in the sky before the passage of a warm front. Cirrus clouds consist of ice crystals which are widely dispersed by strong winds.

climate The weather conditions experienced in any location over the long term. Climate depends upon latitude, altitude and continentality.

clouds Clouds comprise vast quantities of minute water droplets and/or ice crystals. They form when rising air cools to such an extent that it can no longer hold all the water within it as water vapour. The droplets may grow within the cloud—under the right conditions—and precipitation may result. Clouds of varying types can occur. The usual classification recognises cirrus, cirrostratus, cirrocumulus, altocumulus, altostratus, stratus, stratocumulus, nimbostratus, cumulus and cumulonimbus (see these names).

cold front The interface between an advancing cold air mass and a retreating warm air mass. The boundary is angled back over the snout of the advancing cold air. (For an account of the weather conditions upon the passage of a cold front see p. 24.)

computer model A simplified representation of a complex problem projected into the future on a computer, so that the effects of varying one or more than one of the parameters may be observed.

continentality Literally, the effects of distance from the sea on climate. Places located deep within continental interiors tend to be drier because the air masses reaching them have already lost most of their moisture as rainfall or snow. Continentality also has an effect upon temperatures. Water heats up more slowly and loses heat less rapidly than the land. Thus places in mid-latitudes that are in maritime locations tend to have less warm summers and milder winters than places that are in continental interiors.

convection In climate, the upward movement of relatively warm air which has been heated by its contact with the land or the sea.

Coriolis effect The deflection of the wind from a straight path to the right in the northern hemisphere and to the left in the southern hemisphere. This is caused by the rotation of the Earth upon its axis.

cumulonimbus Known as Cb. These huge towering clouds are cumulus clouds that have stretched up so far that they have reached the base of the stratosphere, beyond which they cannot spread. At this point their ice crystals spread out to form the characteristic 'anvil' shape of cumulonimbus. These clouds—which often occur upon the passage of a cold front—bring heavy showers that may be associated with thunderstorms.

cumulus Known as Cu. Detached clouds, which are usually dense but have well-defined outlines. Cumulus clouds have a base which may vary between 460m (1500 ft) and 2000 m (6500 ft). They develop vertically to a considerable height with characteristic domes and towers, and are often seen in the cold sector of a depression after the passage of a cold front. They sometimes spread so far up that they may reach the base of the stratosphere. When a cumulus spreads up to a great height and major 'icing' takes place—often well below the stratosphere—it becomes a cumulonimbus cloud.

cyclic pattern The theory that the climate varies according to a regular or irregular pattern, e.g. the sunspot theory that proposes cycles of 11 and 22 years according to variations of solar activity.

cyclone An area of low pressure; also known as a depression. Within a low pressure area, air is rising and winds blow around the cyclone anticlockwise in the northern hemisphere and clockwise in the southern hemisphere.

deepening The decrease of atmospheric pressure in a depression.

depression See cyclone.

desertification The spread of desert conditions into adjoining semi-arid areas. This process may be caused by climatic factors, either temporary or long-term, or by human activities such as overgrazing, the removal of the natural vegetation or by overcropping and exhausting a poor soil. When desertification occurs the water table drops, there is less surface water and the soil becomes salty. Under such conditions, the natural vegetation of grasses can no longer be supported and the area becomes a desert. The process is best seen in the Sahel of West Africa where the Sahara Desert has advanced since 1968. This has been in part due to a prolonged natural drought but it has been aggravated by overcropping and overgrazing.

dew Precipitation in the form of small droplets deposited on the ground, on plants and on other objects when air close to the surface cools at night to below its dew point.

dew point The temperature at which water vapour in the atmosphere is saturated. When the temperature falls below dew point, water condenses into minute water droplets.

drizzle Fine rain comprising small droplets under 0.5 mm (0.02 in) in diameter.

drought A prolonged period without precipitation. In Britain drought is officially defined as being a period of 15 days or more during which less than 0.2 mm (0.008 in) of rainfall is received.

epoch A measure of geological or climatic time upon the long term scale.

Equatorial climates Those climates experienced in regions near the Equator. They are characterised by high temperatures, high humidity, heavy rainfall and, often, relatively little seasonal variation.

Ferrel cells Wind cells in which air masses rise at the polar fronts in latitudes of 50 to 60 degrees. This rising warm air divides between a weaker return flow towards the Tropics at upper levels and a stronger upper flow towards the Poles. At lower levels winds blow from the Tropics towards the Polar front.

filling An increase of atmospheric pressure in the centre of a depression.

flip A sudden jump or change in climate (see chaos theory).

fog Fog is a cloud which occurs at the Earth's surface. It forms when air is cooled to below its dew point and water vapour condenses around small particles of dust or other particles in the air. This usually occurs when two air masses of differing temperature meet, but can also occur when warm air passes over a cooler surface. The density of the fog depends upon the number of minute water droplets present. Visibility in fog is below 1 km (0.6 mile).

front The interface between two different air masses, as in the warm front and cold front found within a depression.

frost When the air temperature is at or below freezing point, the dew deposited will be in the form of small particles of ice, that is frost.

gale A high wind, on the Beaufort Scale one above force 7.

global warming The gradual warming of the atmosphere of the planet Earth owing to the emission of 'greenhouse' gases, particularly carbon dioxide.

Hadley cells Wind cells immediately North and South of the Equator in which warm air rises and flows away from the Equator at upper levels, to sink again at the Tropics. Winds blow from the Tropics back to the Equator and towards the Poles.

hail A particle of ice which is usually about 5 mm (0.2 in) in diameter, although sometimes may be much bigger. Hail is formed when moist air rises quickly within a cumulonimbus cloud and freezes, then falls rapidly, often rising and falling several times before reaching the ground.

Harmattan A dusty Northeast wind blowing out of the Sahara across the Sahel. It is hot from about March to June and cool from November to February.

high pressure area See anti-cyclone.

humidity The amount of water that is contained within the atmosphere at any given point.

hurricane Fierce tropical storms which originate in the zone between 5 degrees and 20 degrees North and South of the Equator. They begin in areas in which the surface temperature of the oceans is over 27°C (80°F), usually in the summer and the autumn when the seas are at their warmest and intense pressure gradients result. At the centre of a hurricane is an area of very low pressure around which strong winds—of up to 320 km/h (200 miles/h)—may blow. The very centre, or eye, of the storm is calm and experiences only light winds. On the Beaufort Scale (see p. 31) any wind of force 12 or over is classed as a hurricane, and on these grounds the Great Storm of 1987 in Britain was popularly, but incorrectly, termed a hurricane. Strictly speaking hurricanes are only tropical and sub-tropical. In the Pacific, they are usually referred to as typhoons.

isobar An isobar is a line appearing upon a weather map, joining places at which the atmospheric pressure, reduced to sea level, (expressed in millibars) is the same. The closer the isobars appear on a map, the greater the barometric gradient is said to be and the stronger the winds blowing towards the area of low pressure or cyclone.

isohyet A line on a weather map, joining places

that experience the same rainfall for any given time period.

isotherm A line on a weather map, joining places that experience the same temperature for any given time period.

jet stream A strong and enduring wind that flows at high altitude.

latitude An imaginary line on the globe, joining places the same distance North or South of the Equator.

lightning The sudden uprush of air within a cumulonimbus cloud leads to the development of static electricity which is discharged as a flash of lightning.

low cloud Clouds below 2000 m (6000 ft).

low pressure area See cyclone.

maritime climate A climatic type that is influenced by its proximity to the sea; e.g. milder winters owing to the fact that water cools down more slowly than the land does.

Maunder minimum A period between 1645 and 1715 during which very few sunspots were observed. It is named after the British solar scientist Walter Maunder.

Mediterranean climate A climatic type of which the Mediterranean basin is the archetype. It is characterised by hot dry summers and mild wet winters.

microclimate The climate experienced over a very small area, such as a single field or an entire valley.

millibar A unit of pressure used to measure atmospheric pressure.

mist An air mass close to the Earth's surface which has been cooled to below its dew point. The result is condensation into minute water droplets. Visibility in mist is under 2000 m (6000 ft)—thus the difference between fog and mist is basically one of the degree of visibility.

monsoon A period of very wet and frequently stormy weather experienced over much of South Asia between April and December. It is caused by the trade winds being drawn into North India by the development of a deep low over the Thar Desert.

monsoon climate Any climate similar in nature to the monsoon experienced in India and South Asia, e.g. the climate along the coasts of Australia's Northern Territory.

mountain climate Any climate found in mountainous regions that is modified by the effects of altitude.

nimbostratus Known as Ns. A low level cloud whose base varies from 900 m (3000 ft) to 3000 m (10 000 ft). Nimbostratus cloud forms a grey layer which is commonly seen at the passage of a warm front. It invariably brings rain or snow, often continuous and the cloud is thick enough to blot out the Sun.

normal weather See average.

plankton multiplier The effects upon carbon dioxide levels attendant upon a diminution of the plankton in the oceans caused by a rise in sea temperatures. This is thought to be a possible contributor to a 'runaway greenhouse effect'.

polar climate A cold desert climate found in polar regions. What precipitation that does occur comes in the form of snow.

precipitation Water that falls to the ground from clouds. Depending upon circumstances such as temperature it may occur as rain, hail, snow, sleet, dew etc.

pressure gradient See barometric gradient.

prevailing wind The dominant direction from which the wind tends to blow in any given location.

rainfall The total depth of measured precipitation (measured in a rainfall gauge) in any given location—this includes not only rain but snow, hail, dew etc.

rain shadow An area that receives less precipitation because it lies in the lee of mountains over which prevailing moist air masses have to pass and upon which most of their moisture falls. Thus, when the air masses reach the rain shadow, they are too dry to give as much rainfall as on the windward slope. A rain shadow should not be thought of as an area receiving little to no precipitation. Aberdeen is in a rain shadow but it receives an abundant rainfall, although far less than the rainy Grampian Mountains to the West of the city.

ridge of high pressure An extension of higher pressure stretching away from a high.

'runaway greenhouse effect' A vicious circle in which the results of the greenhouse effect cause additional warming which in turn adds to factors thought to be contributing towards global warming.

Sahel, The A broad belt of land stretching across Africa immediately South of the Sahara. It includes parts of Mauritania, Mali, Senegal, Burkina Faso, Niger, Nigeria, Chad and the Sudan and has been characterised since 1968 by drought of varying severity. The Sahel has a savanna grassland climate but is being increasingly encroached upon by desertification.

sleet In Britain, precipitation that is a mixture of both rain and snowflakes. In the USA, it is normally used to describe hail or snow melting as it descends.

snow Precipitation that falls as six-sided feathery crystals of water. It is formed when saturated air condenses at below freezing point. Outside tropical regions most raindrops begin life in clouds as snow.

stratocumulus Known as Sc. A low level cloud whose base from the surface of the Earth varies from 460 m (1500 ft) to 2000 m (6500 ft). These grey or whitish clouds are often seen within a cyclone, indicating that the air has only been able to rise a short distance. Broken, often rounded, these small clouds may, if higher, form a 'mackerel sky'.

stratosphere The part of the atmosphere of the planet Earth that is more than 10 km (6 miles) from the surface.

stratus Known as St. These low level clouds have a base level from the surface of the Earth to 400 m (1500 ft) above the surface. They are uniform grey clouds but the outline of the Sun may be seen through them. Stratus clouds—which may be broken into ragged patches—may give drizzle or light coverings of snow.

subsistence agriculture The type of farming in which the produce is sufficient only to feed the farmer and the farmer's dependents.

sunspot A dark area on the boundary between the Sun's dense interior and its transparent atmosphere. Each sunspot is typically 2000–3000 km (1100 to 1700 miles) across and is darkest at its centre. Sunspots seldom occur individually but rather in groups of two or three or more, and tend to be short-lived, forming and disappearing again within the span of two or three weeks. They represent centres of intense magnetism.

sunspot cycle A cycle of solar activity (lasting 11 years) during which the number of sunspots varies, gradually increasing to a maximum.

temperate climate A climate experienced in mid-latitudes; e.g. temperate maritime climates—which feature warm damp summers and mild wet winters and are experienced in Northwest Europe.

thermal An area of warm rising air.

thermometer A device for measuring temperatures, usually in Celsius (Centigrade).

thunder The noise caused by the explosive effects of the electric discharge that happens when air moves up very rapidly within a cumulonimbus cloud.

tornado A whirlwind circling a small diameter, often of less than 500 m (1600 ft). It is an area of intense strong upward air currents. Tornadoes are capable of causing severe structural and other damage.

trade winds Winds that blow from the areas of high pressure in the tropics to the low pressure area that is a permanent feature in the equatorial region. These winds blow with great constancy. Owing to the rotation of the Earth they are deflected North of the Equator to become the Northeast trades and South of the Equator to become the Southeast trades.

tropics The Tropics of Cancer and Capricorn, respectively 23° 30' N and 23° 30' S of the Equator. They are the limits on the Earth's surface of those areas over which the Sun is directly overhead at least once a year.

troposphere The lower part of the Earth's atmosphere.

typhoon A tropical storm identical in nature to a hurricane. The name is derived from a Chinese word meaning 'great wind'.

warm front The interface between an advancing warm air mass and a mass of cooler air over which it is rising.

water table The level below ground at which the rocks are saturated with water. Where this band of rocks comes to the surface there will be a spring.

weather The day-to-day experience of the climatic elements—rain, wind, sunshine etc.

wind The movement of air from areas of high pressure to areas of low pressure.

Index